FLIGHTS
of FANCY

DEFYING GRAVITY
BY DESIGN & EVOLUTION

Also by Richard Dawkins

The Selfish Gene
The Extended Phenotype
The Blind Watchmaker
River Out of Eden
Climbing Mount Improbable
Unweaving the Rainbow
A Devil's Chaplain
The Ancestor's Tale (with Yan Wong)
The God Delusion
The Greatest Show on Earth
The Magic of Reality (with Dave McKean)
An Appetite for Wonder
Brief Candle in the Dark
Science in the Soul
Outgrowing God
Books do Furnish a Life

RICHARD DAWKINS

FLIGHTS *of* FANCY

DEFYING GRAVITY
BY DESIGN & EVOLUTION

ILLUSTRATED BY

JANA LENZOVÁ

HEAD
of ZEUS

An Apollo Book

First published in the UK in 2021 by Head of Zeus Ltd
An Apollo book

9 7 5 3 2 4 6 8

A catalogue record for this book is available from
the British Library.

ISBN (HB): 9781838937850
ISBN (E): 9781838937874

Designed by Jessie Price
Printed and bound in Slovenia by DZS Grafik

Head of Zeus Ltd
5–8 Hardwick Street
London EC1R 4RG
www.headofzeus.com

For Elon
High-flyer of the imagination

Table of Contents

CHAPTER 1

DREAMS
OF FLYING

'ORNITHOPTER' BY LEONARDO

A scene that happened only in imagination. But WHAT an imagination!

CHAPTER 1

DREAMS OF FLYING

Do you sometimes dream you can fly like a bird? I do and I love it. Gliding effortlessly above the treetops, soaring and swooping, playing and dodging through the third dimension. Computer games and virtual-reality headsets can loft our imagination and fly us through fabled, magical spaces. But it's not the real thing. No wonder some of the past's greatest minds, not least Leonardo da Vinci's, have yearned to join the birds, and designed machines to help them do so. We'll come to some of the old designs later. They didn't work, mostly couldn't have worked, but that didn't kill the dream.

Flights of Fancy means, as you'd expect, that this is a book about flying – all the different ways of defying gravity that have been discovered by humans over the centuries and by other animals over millions of years. But it also includes wandering flights of thought and ideas which take off from thinking about flight itself. Digressions of this kind will be in smaller print, often with the phrase 'by the way...' in bold type.

To begin with fancy at its most fanciful, a 2011 Associated Press poll suggested that 77 percent of Americans believe in angels. Muslims are required to believe in them. Roman Catholics traditionally believe that each of us is looked after by our own private guardian angel. That's a whole lot of wings, beating invisibly and noiselessly around us. According to the legends of *The Arabian Nights*, if you perched on a magic carpet you had only to wish a destination to be instantly whisked there. The mythical King Solomon had a carpet of shining silk, big enough to carry 40,000 of his men. Atop it he could command the winds, and they blew him where he willed. Greek legend tells of Pegasus, a magnificent white horse with wings, which carried the hero Bellerophon on his mission to slay the Chimera monster. Muslims believe the Prophet Muhammad went on a 'night journey' on a flying horse. He hurtled from Mecca to Jerusalem riding the Buraq, a horse-like creature with wings, usually portrayed with a human face like the fabled Greek centaurs. A 'night journey' is something we all experience in our dreams, and some of our dream trips, including flight dreams, are at least as strange as Muhammad's.

The legendary Icarus of Greek mythology had wings made of feathers and wax, linked to his arms. Icarus, in his pride, flew too close to the sun. It melted the wax and he tumbled to his death. A nice warning against getting above yourself, although in reality, of course, he'd have got colder, not hotter, the higher he flew.

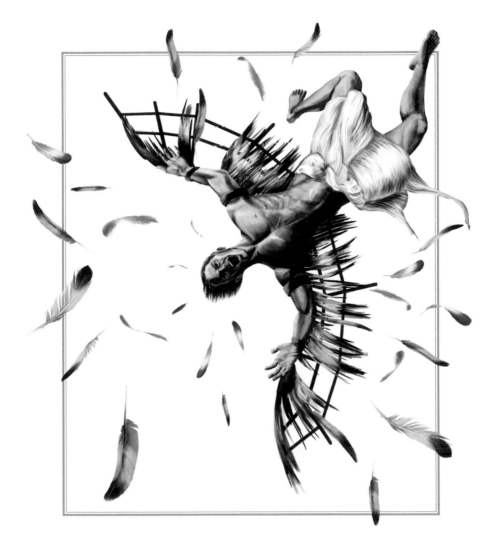

'PRIDE GOETH BEFORE DESTRUCTION
AND A HAUGHTY SPIRIT BEFORE A FALL'
Icarus flew too close to the sun and fell to his death.

CONAN DOYLE BELIEVED IN FAIRIES
Neither Sherlock Holmes nor Professor Challenger would have fallen for
the hoax that fooled their creator. But he was a wonderful writer!

Witches were supposed to whizz through the air on broomsticks, and Harry Potter has recently joined them. Santa Claus and his reindeer speed from chimney to chimney high above the December snow. Meditating gurus and fakirs fake their claim to hover above the floor in the lotus position. Levitation is a myth popular enough to inspire cartoon jokes, nearly as many as desert island jokes. My favourite, unsurprisingly, is from the *New Yorker*. Man in street looks at door high up in wall. Label on door: 'National Levitation Society'.

Sir Arthur Conan Doyle created the forensically rational Sherlock Holmes, first among fictional detectives. Another of Doyle's characters was the formidable Professor Challenger, a ferociously rational scientist. Doyle evidently admired both, yet he allowed himself to be fooled by a childish hoax in a way his two heroes would have scorned. Literally childish, for he was gulled by a pair of playful children who made trick photographs of winged 'fairies'. Two cousins, Elsie Wright and Frances Griffiths, cut out pictures of fairies from a book, stuck them on cardboard, hung them up in the garden and photographed each other hanging out with them. Doyle was only the most famous of many people fooled by the 'Cottingley Fairies' hoax. He even wrote a whole book, *The Coming of the Fairies*, pushing his strong belief in those little winged people flitting like butterflies from flower to flower.

The irascible Professor Challenger might have roared the question, 'From what ancestors did the fairies evolve? Did they arise from apes independently from ordinary humans? What was the evolutionary origin of their wings?' Doyle himself, as a doctor knowing some anatomy, should have wondered whether fairy wings evolved as projections of the shoulder blades, the ribs, or something completely novel. For us today, the photographs look obviously faked. But to be fair to Sir Arthur, this was long before Photoshop and it was widely believed that 'the camera cannot lie'. We of the Internet-savvy generation know that photos are all too easy to fake. The 'Cottingley' cousins eventually admitted to their prank, but not till they were over seventy and Conan Doyle long dead.

The dream continues. It lifts our imagination every day as we fly through the Internet. As I type these words in England they 'fly up' into the Cloud ready to come 'down' to an American computer. I can log in to an image of the spinning world and 'fly' virtually from Oxford to Australia, looking 'down' on the Alps and the Himalayas on the way. I don't know whether the anti-gravity machines of science fiction will ever become real. I doubt it and will mention the possibility no further. Without straying from science fact, this book will list the ways in which gravity can be tamed, though not literally escaped. How have humans, with our technology, and other animals, with their biology, solved the problem of rising above the solid ground: escaping, if only temporarily or partially, the tyranny of gravity? But first we need to ask why it might be a good thing for animals to get off the ground in the first place. What, in the world of nature, is flight good for?

CHAPTER 2

WHAT IS
FLIGHT
GOOD FOR?

CHAPTER 2

WHAT IS FLIGHT GOOD FOR?

There are so many ways to answer this question, you may wonder why we should even bother to ask. We have to go beyond dreams of blissful floating among mythical clouds, and – forgive me – come down to earth. We have to give a precise answer. And for living organisms that means a Darwinian answer. Evolutionary change is how all living creatures got to be the way they are. And, where living creatures are concerned, the solution to every 'What is it good for?' question is always and without exception the same: Darwinian natural selection or 'survival of the fittest'.

What then, in Darwin language, are wings 'good' for? Good for the animal's survival? Yes, of course, and we'll come soon to the many particular ways in which that answer plays itself out in practice. For example, spotting food from above. But survival is only part of the story. In a Darwinian world, survival is only a means to the end of reproduction. Male moths typically use their wings to surf the breeze towards a female, guided by her scent – some can detect it even if it's diluted to one part in a quadrillion.

'I SMELL A FEMALE 3 MILES AWAY'
*Antennae like the feathery beauties on this moth can sense a female on
the breeze from a huge distance. Male moths fan air over their antennae,
turning as they do so: scanning round all points of the compass.*

They do it by means of their huge and highly sensitive antennae.
It doesn't help the male's own survival but, as I said, survival is
only a means to the end of reproduction.

We can refine that statement yet further, and in doing so we return to the idea of survival. Survival, not of individuals but of genes. Individuals die but genes live on as copies. The kind of survival that is achieved by reproduction is the survival of genes. Genes, the 'good' ones anyway, survive through many generations, even millions of years, in the form of faithful copies. The bad ones don't survive – that's what 'bad' means if you are a gene. And how does a gene qualify to be 'good'? Good at building bodies that are good at surviving, reproducing and passing on those very same genes. Genes for making giant antennae on moths survive because they pass into eggs laid by females that those antennae detect.

In the same way, wings are good for the long-term survival of genes *for* making wings. Genes for making good wings helped their possessors to pass on those very genes to the next generation. And the next. And so on until, after countless generations, what we see is animals that fly very well indeed. In recent times (recent by evolution's standards) human engineers have rediscovered how to fly – in similar ways to animals, which is not surprising because physics is physics, and evolving birds and bats had to battle with the same physics as human aircraft designers today. But while planes really *are* designed, birds and bats, moths and pterosaurs were never designed but were shaped by the natural selection of their ancestors. They fly well because, through past generations, their ancestors flew slightly better than poorly flying rivals who therefore failed to become ancestors – and failed to pass on the genes for flying poorly. I've explained all that more fully in other books, but the previous paragraph and this one will suffice here before we turn to the detail of what flying is good for. And that varies from species to species. As we shall now see.

Some birds, like peacocks for whom flying is a big effort, lift their bulk into the air for a short distance to escape predators, then come down a safe way away. Flying fish do the same in the sea. Flight in these cases could be seen as an assisted jump. Many birds, not just poor flyers like peacocks, use flight to evade predators who are stuck on the ground. And, of course, some predators are not stuck on the ground: they can fly, too. An aerial arms race develops over evolutionary time. Prey get faster to escape capture, and predators get faster in reply. Prey evolve twist-and-turn evasive manoeuvres, and predators evolve counter-moves in return. A beautiful example is the arms race between night-flying moths and the bats that prey on them.

Bats find their way through the dark, and home in on their prey, using a sense that we can hardly imagine. Their brains analyse echoes from their own ultrasonic (too high-pitched for us to hear) sound pulses. As a bat comes within range of a moth, it increases the slow tick... tick... tick base-rate of sound pulses to a rapid-fire rat-a-tat-tat, then to a brrrrrrrrrrrrrrrrrrrrrrrr in the final attack phase. If you think of each sound pulse as sampling the world, you can easily see why increasing the sampling frequency would improve the accuracy of pinpointing the target. Over millions of years, evolution perfected the bats' echo technology, including the sophisticated brain software that served it. At the same time the moths, on the other side of the arms race, were doing some smart evolving of their own. They evolved ears tuned to just the right ultra-high pitch to hear bat shrieks. They evolved unconscious, automatic evasive tactics that come into play whenever they hear a bat: swooping, diving, dodging. And in their turn the bats responded by evolving faster reflexes and nimbler flying skills. What we see at the climax of the arms race looks like the legendary aerial dogfights between the Spitfires and Messerschmitts of the Second World War. The drama takes place at night in what can seem to us like total silence because our ears, unlike those of the moths, are deaf to the machine-gun pulses of the bats. The moth ears are tuned to little else. Bats are probably the main reason they have ears at all.

⤳ **By the way**, protection from bats may also be one reason why moths are furry. Acoustic engineers, wishing to reduce echoes in a room, line the walls with material that has sound-absorption properties similar to moth fur. But some moths have an additional, more ingenious trick. Their wings are covered with tiny forked scales tuned to resonate with bat ultrasound in such a way as to 'disappear off the radar' like a stealth bomber. Some moths make ultrasonic noises themselves, which may serve to 'jam' the bats' radar (strictly sonar). And a few species of moth use ultrasound for courtship.

Birds that forage on the ground use flight to travel quickly from one foraging area to another as each becomes depleted. Vultures and birds of prey use their wings to gain a high vantage point to scan for food over a large area. Vultures do it from very high indeed. Their prey is already dead, they don't have to hurry to catch it, so they can afford to go really high and can scan a huge area for tell-tale signs of, say, a lion kill. Often those tell-tale signs include other vultures. Having spotted a corpse, they then glide down. Raptors, like eagles and hawks, seeking live prey, search from lower altitudes then swoop, often at great speed. Many fishing birds such as terns and gannets do something similar, in the technique known as plunge-diving.

Gannets scan large areas of open ocean for tell-tale signs of fish schools, perhaps a darkening of the surface, or the sight of other birds already there. A dense flock of gannets, or the closely related boobies, storming down from a height, bombarding a shoal of fish at 60 miles per hour, is one of the great sights life has to offer. Their relentless blitzkrieg recalls another Second World War image, the Stuka dive-bombers with their screaming 'Jericho Trumpets', or the Japanese kamikaze planes. Except that gannets and boobies are not hurtling to their deaths. Not normally, but a misjudged dive can break their neck. And over a long period, a lifetime of plunge-diving does progressive damage to their eyes: the life of a booby may eventually be brought to its end by poor eyesight. You could say they are shortening their lives by diving. But they'd shorten them even more by not diving, because they

then might starve. Gannets are such narrowly specialist divers, if they lose that skill they can't compete against other birds such as gulls foraging on the surface.

> ↪ **By the way**, there's an interesting lesson in evolutionary theory here, one that will keep cropping up throughout this book: it's the lesson of compromise. Darwinian natural selection may drive an animal to shorten its life when old if, in doing so, it increases its success at reproducing when young. As we've just seen, 'success', in Darwinian language, precisely means leaving behind lots of copies of your genes before you die. Genes that drive a gannet to fish more efficiently when young succeed in getting passed on to the next generation even if they also hasten the bird's death when old. This kind of reasoning can help us understand why we age, even though we don't plunge-dive for fish. We inherit the genes of a long line of ancestors that were good at being young. They didn't have to be good at being old: they'd mostly done their reproducing by then.

Gannets are fast, but the champion dive-bombers are falcons, who catch birds on the wing. During a prey-catching dive or *stoop*,

---

◁ THE STUKA DIVE-BOMBERS OF THE BIRD WORLD
Gannets and boobies are master fishers from the air.
Only one gannet is shown here, but a large flock dive-bombing
together is a sight never to be forgotten.

a peregrine falcon can reach an astonishing 200 miles per hour. To dive through the air at 200 miles per hour, the best shape to be is quite different from the best shape when in level flight, scanning for prey. Sure enough, the diving peregrine folds its wings like a swing-wing fighter jet. Such colossal speeds present problems and dangers. The birds would be unable to breathe without their specially modified nostrils (whose design has been partially copied in the jet engines of very fast planes). A botched impact at such a breakneck speed could literally break the bird's neck. As with the gannets, there are no doubt trade-offs between short-term benefits in reproductive success on the one hand, and risk of shortening life on the other.

What else is flight good for? Cliff ledges are excellent for nesting and roosting, safe from ground predators like foxes. Kittiwakes are gulls that specialise in building their nests on ledges so precariously inaccessible that predators – even other flying birds – have a hard time raiding them. Many birds seek safety for their nests in trees. Wings provide a wonderfully fast way to get up in a tree, and to carry grass and other nest material and, later, food for chicks. Many trees are covered with fruit: food for toucans, parrots, numerous other birds and the larger species of bat. To be sure, monkeys and apes also harvest fruit up trees by climbing, but even the most athletic monkey or ape is no match for a bird in a race through the branches. Gibbons are the most athletic of all tree climbers, and they have perfected a technique called *brachiation*, which is almost like flying.

THE CLIMAX OF AN EVOLUTIONARY ARMS RACE
Peregrine falcons can dive onto flying prey (the other side of the arms race) at up to 200 miles per hour.

Brachiating (from *brachium*, the Latin for 'arm') means swinging through the trees, using very long arms almost as though they are legs running upside down through the air. A gibbon in full flight – I use the word almost literally – hurtles through the canopy at breathtaking speed, flinging itself from one branch to the next which may be many metres away. Not flying in the strict sense, but it almost amounts to the same thing. Our ancestors probably brachiated at one stage in our history, though I'm sure we could never have outpaced a gibbon.

Flowers manufacture nectar, which is the main aviation fuel of hummingbirds and sunbirds, butterflies and bees. Bees feed their larvae on pollen, which they also gather from flowers. The whole family of bees, within the wider class of insects, is dependent on flowering plants, and they evolved together ('co-evolved'), beginning about 130 million years ago in the Cretaceous period. What better way than on wings, to move swiftly from flower to flower?

Most insects fly, and catching them on the wing has become a fine art for swallows and swifts, flycatchers and smaller species of bats. Dragonflies, too, skilfully catch insects on the wing, using their large eyes to spot them.

Swifts are specialist insect eaters and they catch them entirely in flight. They have taken life in the air to such an extreme that they almost never set foot on land. They even achieve the difficult feat of mating in the air. Just as turtles left the land for a watery life, ancestral swifts left the land for the air. Both return

WHOLE LIFE ON THE WING
Swifts have pushed life on the wing to the limit. They even mate without landing. Does walking on land feel as alien to them as underwater swimming does to us?

only to lay eggs. And, in the swifts' case, to incubate them and feed the chicks. You get the feeling that, if only it were possible to lay eggs on the wing, swifts would do it, just as whales have gone one better than turtles and never return to land for any reason at all.

As their name suggests, swifts are extremely fast, and they remind us that great speed of travel is a major advantage of flight. A century ago, the great ocean liners took many days to cross the

Atlantic. Nowadays we fly it in hours. The difference is mainly due to water's greater friction compared with air. Even in air it varies with height. The higher a plane flies, the lower is the drag in the rarefied air, which is why modern airliners fly as high as they do. Why don't they fly even higher? For one thing, they'd be starved of the oxygen their engines need for combustion of the fuel. Rocket motors designed to work outside Earth's atmosphere carry their own oxygen. Other things affect the design of planes that fly at very high altitudes. As we'll see in Chapter 8, they need air to obtain lift, and at very high altitudes the air is so thin that they need to fly faster to get their lift. Planes designed for low altitudes don't perform well at high altitude, and vice versa. Rockets don't need air for lift, and they don't need wings. Their engines propel them directly against gravity. And, once they have achieved orbital velocity, they can switch off and float weightlessly while still travelling extremely fast.

As a child I used to worry that rocket motors wouldn't work in outer space because there is no air behind them to 'push against'. I was wrong. 'Pushing against' has nothing to do with it. Here's why. First, a couple of more down-to-earth parallels. When a big artillery gun fires, there's a huge recoil. The entire gun rocks back on its wheels as the shell emerges from the barrel. Nobody thinks the recoil is caused by the shell 'pushing' against the air in front of the gun. What really happens is this. There's an explosion inside the cartridge of the shell. Gas pushes violently in all directions. The sideways forces cancel each other out. The forward force pushes the shell out through the barrel, meeting little resistance. The backward force presses against the gun, rocking it backwards on its wheels. The same recoil would enable you, sitting on a toboggan, to propel yourself over the ice. You'd fire a rifle in a direction opposite to the direction you want to go. If you're interested in physics you'll know that what we have here is Newton's Third Law: 'To every action there is an equal and opposite reaction.' It's not because the bullets are pushing against the air that the toboggan moves. You'd travel even faster in a vacuum. And the same goes for a rocket engine in a vacuum.

The tilt of Earth's axis means there are changing seasons as the planet races around the sun. This means that the best place to be, for feeding or breeding, changes from month to month. For many animals, the cost of moving a great distance is outweighed by the benefit of finding better weather, with all its consequences. And of course, 'better' may not mean what we humans think of as fine weather, good for a summer holiday. Whales migrate from warm breeding grounds to cooler waters where currents provide a rich upwelling of nutrients to supply the food chain on which they depend. Wings enable birds to cover huge distances. Many species of bird migrate but the distance record is held by the Arctic tern, which flies all the way – 12,000 miles – from the

Arctic Circle, where they breed, to the Antarctic Circle and back every year. The journey takes them only two months. Prodigious distances on this scale and in such a short time could be covered only in the air. Arctic terns get two summers with no winters each year, and this extreme example gives the clue to why so many animals migrate.

Many migrating animals, not just birds, show great feats of accurate navigation as well as great feats of endurance. Our European swallows winter in Africa, then return the following summer to the identical spot, their very own nest – a staggering feat of precision navigation. How birds do this kind of thing has long been a mystery. It's now in the process of being solved. Ornithologists label individual birds with tiny leg rings (American 'bands') and nowadays tiny GPS transmitters, so they can plot their movements. Even radar has been used to track the course of great flocks of migrating birds. We are starting to understand that birds use several navigation techniques, different combinations of methods being preferred by different species and at different stages in the migration process.

Familiar landmarks are part of the story, certainly in the final stages of the journey when migrant birds return to last year's nest. But during the course of the long journey, too, birds are known to follow rivers, coastlines or mountain ranges. In many species, young birds on their first migration need to be accompanied by older, experienced birds who know the geography. As well as landmarks, birds are often assisted by built-in compasses.

WORLD RECORD LONG-
DISTANCE MIGRANT
*Migrating from pole to pole, the Arctic
tern never sees a winter – only polar
summers, 12,000 miles apart.*

It is now established that some species are sensitive to Earth's
magnetic field. It isn't always clear how they see or feel the
compass direction, but it has been shown that they do. And 'see'
may well be the correct verb to use, for a leading theory of the
mechanism, whatever it is, locates it in the eye.

It has long been known that migrating birds (along with

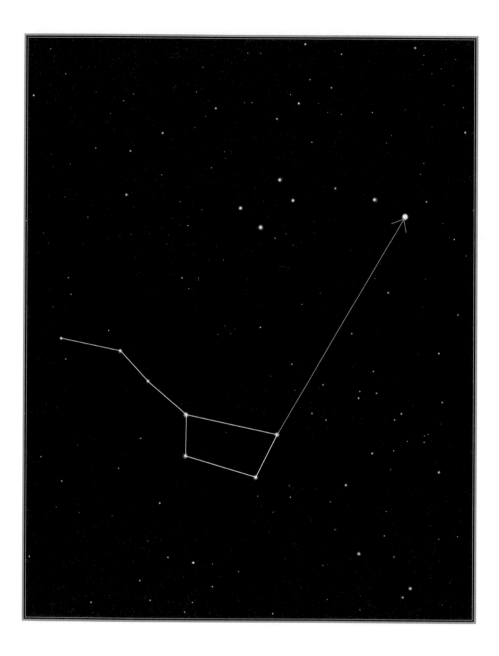

'ALL I ASK IS A TALL SHIP AND A
STAR TO STEER HER BY'

*Just draw an imaginary line upwards through the two stars most distant
from the handle of the Big Dipper (the 'pointers') and continue out to the
first bright star you hit. That will be Polaris, the North Star.*

insects and other animals) also use the sun as a compass. Of course, the sun changes its apparent position from east in the morning to west in the evening via south (or north if you are in the Southern Hemisphere) at noon. This means that a migrating bird can use the sun as a compass only if it also knows the time of day. And all animals do have an internal clock. Indeed, each cell has a clock. Our internal clock is what makes us want to do particular things, or feel hungry or sleepy, at regular times of the day or night. Researchers have experimentally put people in underground bunkers totally cut off from the outside world. They continue to perform all their normal activities, sleeping and waking, switching the light on and off, taking meals, etc., with a 24-hour rhythm. As you'd expect, it isn't exactly 24 hours – it might be 10 minutes longer, for instance – so it gradually strays out of step with the outside world. That's why it's called a 'circadian' cycle (*circa* is Latin for 'around') instead of simply a 'dian' (*dies* is Latin for 'day') cycle. Under normal conditions the circadian clock is reset by sight of the actual sun. Migrating birds, like all animals, are equipped with a clock such as they would need if they are to use the sun as a compass.

Some migrators fly at night, so they can't use the sun. They can use the stars instead. Most people know that one particular star, Polaris, is almost directly above our North Pole, regardless of Earth's spinning. In the Northern Hemisphere, therefore, Polaris can reliably be used as a compass. But how do you know which of many stars is Polaris? When my sister and I were little, our father taught us a great number of useful things. Among them he showed us that you can find Polaris from the easily recognised Plough or Big Dipper (part of the Great Bear or Ursa Major constellation). Just draw an imaginary line upwards through the two stars most distant from the handle of the Big Dipper (the 'pointers') and continue out to the first bright star you hit. This is Polaris, and at night you can use it to steer. If you are in the Northern Hemisphere, that is. If, like the Polynesians navigating among remote Pacific islands, you're in the Southern Hemisphere you have to be a bit more sophisticated: there's no convenient bright star hovering over the South Pole. The Southern Cross constellation is nowhere near close enough. We'll return to this problem.

Even in the Northern Hemisphere, where Polaris is obligingly available, how do night-flying birds know which star is which, for navigation purposes? Theoretically they could inherit a star map in their genes, but that seems a bit far-fetched. There's a more plausible way, and we know it's true for North American indigo buntings because of a brilliant series of experiments done in a planetarium by Stephen Emlen of Cornell University.

☞ Indigo buntings are a beautiful blue colour and could rightfully be called 'bluebirds', unlike anything we have in Britain – notwithstanding the weird reference to bluebirds in the song 'English Country Garden', with its joyfully prancing tune arranged by the Australian composer Percy

Grainger. (By the way, there really are some gorgeously blue birds in Australia.) There's also a patriotic wartime song that says, 'There'll be blue birds over the white cliffs of Dover'. It would be nice if this were a poetic reference to the blue uniforms of the Royal Air Force – 'the Few' – but perhaps it's just that the American poet didn't realise there are no blue birds in Britain. Or maybe it was 'poetic licence' – and nothing wrong with that!

Indigo buntings are long-distance migrants and they fly at night. During the migration season, caged buntings flutter against the bars on the side towards which they would normally migrate. Dr Emlen devised a method for measuring this thwarted preference, using a special circular cage. The lower part of the cage was a cone, a funnel lined with white paper, with an ink pad at the bottom on which the birds often alighted. The birds fluttered up the cone and their inky footmarks on the paper recorded their preferred direction. This apparatus has since been used by other researchers on bird migration, and has become known as the 'Emlen Funnel'. The buntings' preferred direction in autumn was broadly south, corresponding to their normal migration to their wintering grounds in Mexico and the Caribbean. In spring, they fluttered closer to the north side of the Emlen Funnel, corresponding to their normal return journey to Canada and northern America.

CONSTANT AS THE NORTH STAR?
The indigo bunting's inky footmarks on the side of the Emlen Funnel
indicate the direction in which it 'wants' to migrate (not to scale).

Emlen was in the fortunate position of being able to borrow
a planetarium, and he put his funnel cage in it. He did a series
of fascinating experiments manipulating the artificial star map,
blotting out selected patches of the artificial sky, and so on. In this
way he was able to prove that indigo buntings do indeed use the
stars to steer, especially stars in the vicinity of Polaris, including
those of constellations such as the Big Dipper, Cepheus and
Cassiopeia (these are Northern Hemisphere birds, remember).

Perhaps the most interesting of his planetarium experiments was the one in which Emlen asked the question, 'How do the birds know which stars to use for navigation?' Rather than appealing to a genetic star map, his hypothesis was that young birds, before migrating, take time to watch the rotating sky during the night, and learn that a certain portion scarcely rotates at all because its stars are close to the centre of rotation. This method would work even if Polaris didn't exist: there'd still be a patch of sky which could be recognised as not rotating, and that would be north. Or south if you are a Southern Hemisphere bird.

Emlen tested this idea with a most ingenious experiment. He hand-reared baby birds and exposed them, as they grew up, only to planetarium stars. Some of the birds were given experience of a planetarium night sky which rotated about Polaris. When tested in the funnel cage in autumn, they showed a preference for the normal migration direction. But another group of young birds was reared under a different condition. Again the only stars they saw while growing up were planetarium stars. But in this case Emlen cunningly manipulated the planetarium so that its night sky rotated not about Polaris but about Betelgeuse, another bright star (you can recognise it as Orion's left shoulder if you live in the Northern Hemisphere, his right foot if you live in the south). What did those birds do when they were eventually tested in the funnel cage? Wonderful to report, they treated Betelgeuse as if it were due north, and used it to steer in a false direction.

But now we need to make a distinction between 'map' and 'compass'. To fly, say, south-west, you need only a compass. But for a homing pigeon a compass is not enough. Homers need a 'map' as well. They are shut away in a basket, transported far in a random direction, and released. They fly home so quickly that they must have some means of telling where they have been released. Moreover, experimenters with homing pigeons don't record only whether the birds get home safely. In many cases they also, after uncaging a bird at the release point, follow the bird with binoculars, noting its compass direction at the moment when it disappears from view. Homing pigeons show a marked tendency to disappear in the direction of home, even though they are too far away to be using familiar landmarks.

Before radio, armies used homing pigeons to carry messages back to headquarters. The British army in the First World War used a modified London bus as a field dovecot. In the Second World War, the Germans deployed specially trained hawks to intercept British messenger pigeons. This set off an ornithological arms race, with British special agents licensed to kill the hawks.

A compass, then, however accurate, is not enough for a homing pigeon. Before it can even begin to use its compass, the homer has to know where it is. It's not just homing pigeons: any long-distance migrant also might need a map to compensate for being blown off course. Indeed, experimenters have artificially 'blown' migrating birds off course – caught them in mid migration and then released them from a displaced location: moved them, say,

'I KNOW WHERE I AM, AND I KNOW WHERE I'M GOING'
Homing pigeons need a map as well as a compass.

100 miles east, and then released them. Instead of proceeding in the same compass direction, which would have landed them 100 miles east of where they ought to be, the birds managed to end up at their correct destination. Compensating for being blown off course could be how 'homing' originally evolved in birds, long before humans invented baskets and cars or trains to transport them.

Various theories of bird 'maps' have been proposed. Undoubtedly for experienced birds, familiar landmarks play a role. There is evidence that smells, which I suppose you could call a special kind of landmark, are important. A theoretical but probably impractical possibility is *inertial navigation*. Sitting in a car, even if you are blindfolded, you can detect acceleration and deceleration (though not uniform motion, as Einstein reminded us), including changes in direction. Theoretically, a pigeon sitting in its dark basket could add up all the speed-ups and slow-downs, all the left turns and right turns, as the car transports it from home loft to release point. The bird could theoretically then calculate where the release point is in relation to the home loft.

An experimenter called Geoffrey Matthews tested the inertial navigation theory. He put his pigeons in a cylindrical lightproof drum, continuously rotating as he drove them from their home loft to the release point. Even after such ruthless treatment the poor creatures managed to set a true course for home. To say the least, this renders the inertial navigation hypothesis improbable. I need to correct a wrong, here. In a popular book, this experimental apparatus was alleged to be one of those mobile cement mixers you see churning round and round on the back of a lorry. That's a vivid image in keeping with Dr Matthews's sense of humour, but it isn't true.

Humans can calculate where they are from astronomical measurements. Seafarers have long used sextants to pinpoint their position. In the Second World War my father's brother, forbidden for security reasons to know the location of the troopship he was on, cleverly – and typically – made himself a sextant in order to discover that very thing. He was almost arrested as a spy. A sextant is an instrument for measuring the angle between two targets, for example the sun and the horizon. You can use that angle at local noon to work out your latitude, but you have to know when it is noon locally, and that varies with longitude. If you have an accurate clock which tells you what time it is at some reference longitude such as the Greenwich Meridian (or your home loft if you are a pigeon) you can compare it with local time, and that theoretically gives you your longitude. But again, how can they know what time it is locally? The same Geoffrey Matthews suggested that birds observe not just the height of the sun but also the arc movement of the sun over a period. Of course they'd have to watch the sun for some little

DID SAILORS REDISCOVER

BIRD TECHNOLOGY?

Could homing pigeons be using
something equivalent to a sailor's
sextant? It's not a silly idea but
more evidence is needed.

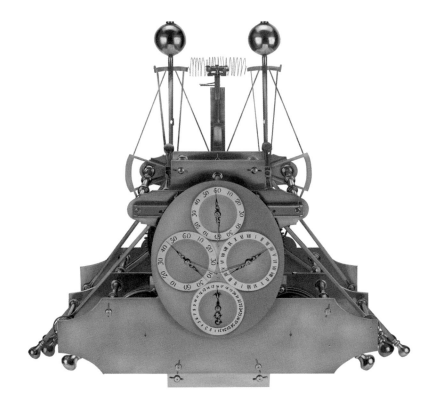

HARRISON IMPROVED MARINE CHRONOMETER
Such intricacy of parts, such finely honed complexity, each tiny
improvement shaving off a few more miles of potentially fatal error
in navigation. Migrating birds don't need quite the same precision
(they don't get shipwrecked), but how do they do it anyway?

time in order to extrapolate the arc. That may seem improbable, but we know from Emlen's planetarium experiments that young indigo buntings can do something not altogether different as they notice which part of the sky is the centre of rotation. And Matthews's student Andrew Whiten did laboratory experiments on pigeons which showed that they are capable of the necessary feats of discrimination.

By extrapolating the arc of its apparent movement, pigeons could theoretically work out where the sun will be (or was) at its highest point, the zenith, at local noon. We've already seen that the height of the sun at the zenith tells them their latitude. And the horizontal angular distance from the calculated zenith tells them the local time. If they compared this local time with time at the home loft (their private Greenwich), given by their internal clock, this could give them their longitude.

Unfortunately, even a small inaccuracy in the clock makes for a big error in navigation. The celebrated mariner Ferdinand Magellan, on the first ever voyage around the world, carried eighteen hourglasses. If he had used these for navigation, the error would have been huge. This was still a problem in the eighteenth century, which is why the British government announced a competition, with large prize money, for the invention of a marine chronometer, a precision clock that would remain accurate in spite of the rolling of the sea – as pendulum clocks would not. The prize was won by John Harrison, a Yorkshire carpenter. While it's true that homing pigeons, like all animals, have an internal clock, it's no match for the Harrison chronometer, or even Magellan hourglasses. On the other hand, flying birds perhaps don't need the same accuracy as sailors, who could shipwreck if they got their navigation wrong. Other astronomical theories of the same general type as the Matthews hypothesis have been proposed to solve the riddle of long-distance navigation by birds.

What other maps might birds use? Magnetism-based maps are possible and sharks are known to use them. Different locations on the Earth's surface have their own characteristic magnetic signature. What might such a signature look like? Here's one possible idea for the kind of thing it might be. This theory makes use of the fact that the direction of magnetic north (or south) is not exactly the same as true north (or south). A magnetic compass is measuring the Earth's magnetic field which is only approximately aligned with the axis of the planet's spinning. The discrepancy between magnetic north and true north is called the magnetic declination, and all compass users who need accuracy must take account of it. The declination varies from place to place (and from time to time because of shifts of the Earth's core, which can even flip the Earth's magnetic field on its head now and again as the centuries go by). If you can measure the declination, for example by measuring the angle between Polaris and the north-pointing needle of a magnetic compass, you can work out (also using the intensity of the magnetic field) where you are. That could be the magnetic signature we are looking for.

There is some extraordinary evidence that Russian reed warblers can do this. Using birds in Emlen Funnels, experimenters artificially shifted the magnetic field by 8.5 degrees. If the birds were simply using a magnetic compass, their preferred direction in the Emlen Funnel should have shifted through the same angle, 8.5 degrees. In fact, they shifted their fluttering direction through a massive 151 degrees. The 8.5 degree shift in magnetic field, when fed into a calculation based on declination, would tell them they were no longer in Russia but in Aberdeen! And, lo and behold, the direction they preferred in the Emlen Funnel was the direction they *would* need to take if for some reason they found themselves in Aberdeen and were seeking to end up in their normal migration destination. That Aberdeen signature is an example of the kind of thing a magnetic signature might look like. It's a step in the direction of understanding how a magnetic sense might amount to more than just a compass. I must admit it sounds to me a bit too good to be plausible.

Needless to say, nobody is suggesting that birds consciously do sophisticated calculations such as would be required for the Matthews theory of sun navigation. Of course, the birds have no equivalent of pencil and paper, no tables of magnetic declination or field strength. When you catch a ball in the deep field at cricket or baseball, your brain is performing the equivalent of solving sophisticated differential equations. But you are unaware of this as you control your legs, eyes and grasping hands to make the catch. Same with the birds.

Animals with wings can get to islands that mere legs can't reach. Remote islands often have no mammals. Or the only mammals (except for human introductions like dingoes or stowaway rats) are bats. Why bats? Because bats have wings, of course. Apart from bats, remote islands largely belong to birds rather than mammals. On islands, we commonly find that the ground trades normally plied by mammals are instead monopolised by birds. New Zealand's national bird, the kiwi, makes its living like a ground-dwelling mammal. Its ancestors could fly, which is presumably how they managed to get there originally. Kiwis are typical of island birds in having shrunk their wings so they can no longer fly, as we shall see in the next chapter. But wings are the reason they arrived on an island in the first place.

An island bird's flying ancestors arrived by accident, perhaps blown off course by a freak wind. And here I need to stress a difficult, subtle point. This chapter is about what flight is good for. Finding food, escaping from predators, migrating every year to summer feeding grounds, these are all straightforward benefits of wings. Natural selection perfected the wings – every detail of their shape and workings – for the benefit of the birds doing the flying. Having the luck to colonise a remote island is different. Wings are not shaped by natural selection for the purpose of finding islands to colonise and evolve in. If wings are a benefit here, we are using 'benefit' in a rather peculiar sense. We're talking here about a rare, freak event. Perhaps a catastrophic hurricane blew a migrating, egg-bearing female off course and dumped her on the island as a lucky accident.

Even wingless mammals occasionally get dumped on islands by freak accidents. Nobody knows how rodents and monkeys arrived in South America. In both cases it happened about 40 million years ago and the result has been a glorious profusion of different types of monkeys, and rodents – relatives of guinea pigs. The map of the world looked different 40 million years ago. Africa was closer to South America, and there were islands between them. Monkeys and rodents probably island-hopped, perhaps floating on rafts of vegetation, or trees swept out to sea by hurricanes. Such freak events only had to happen once, after which the newly arrived castaways found a nice new place to live, breed and – eventually – evolve. Same thing with birds except that wings gave them a start-up edge. Even so, we'd be wrong to say that enjoying such freak colonisations was a benefit of having wings; not a benefit in the same conventional sense as gaining height to spot food every day is a benefit of wings.

Flying would seem to be a hugely useful ability, useful for all sorts of purposes. You might ask, then, why don't all animals fly? More sharply phrased, why have many animals gone so far as to lose the perfectly good wings their ancestors once had?

IF FLYING IS SO GREAT, WHY DO SOME ANIMALS LOSE THEIR WINGS?

PIGS MIGHT FLY

They don't, but could they ever? If not, why not? When is
it ever sensible to wonder why animals don't do something?
Like asking why do some animals not fly?

CHAPTER 3

IF FLYING IS SO GREAT, WHY DO SOME ANIMALS LOSE THEIR WINGS?

> And why the sea is boiling hot —
> And whether pigs have wings.
> Lewis Carroll, *Through the Looking-Glass*, 1871

The sea is not boiling hot, though one day (about 5 billion years hence) it will be. And pigs certainly don't have wings but it's actually not a silly question to ask why not. It's a kind of jokey approach to a more general question: 'If such and such is so great, why don't all animals have such and such? Why don't all animals, even pigs, have wings?' Many biologists would say, 'It's because the necessary genetic variation to evolve wings was never available for natural selection to work on. The right mutations didn't arise, and perhaps couldn't because pig embryology is simply not geared to sprout little projections which might eventually grow into wings.' I am perhaps eccentric among biologists in not leaping immediately to that answer. I would add a combination of the following three answers: 'Because wings wouldn't be useful to them; because wings would be a handicap in their particular way of life; and because even if wings might be useful to them, the usefulness would be outweighed by the economic costs.' The fact that wings are not always a good thing is demonstrated by those animals whose ancestors used to have wings but who have given them up. That's what this chapter is about.

Worker ants don't have wings. They walk everywhere. Well, perhaps 'run' is a better word. The ancestors of ants were winged wasps, so modern ants have lost their wings over evolutionary time. But we don't have to go back that far. Nowhere near. The worker ant's immediate parents, her mother and her father both had wings. Every worker ant is a sterile female fully equipped with the genes of a queen, and she would have sprouted wings if she had been reared differently, as queens are. The potential for wings is, so to speak, coiled up in the genes of all ants but in workers it doesn't burst forth. There must be something wrong with having wings, otherwise worker ants would realise their undoubted genetic ability to grow them. The pluses and minuses for and against wings must be pretty finely balanced if a female sometimes grows them and sometimes doesn't.

Queens need their wings to found a new nest far from their original home nest. We'll come to why this would be a good thing in Chapter 11. Wings also enable young queens to meet winged males not from their own nest. Again, we'll see later why

QUEEN ANT SHEDDING HER NOW USELESS WINGS
*A worker ant never develops wings even though both her parents
did, and her genes know exactly how to make them. Wings are not
all they're cracked up to be.*

such outbreeding might be a good thing. Workers, since they
don't reproduce, have neither of these two needs. They typically
spend a great deal of their time underground, crawling through
confined spaces. Perhaps wings would get in the way in the
cramped corridors, galleries and chambers of an underground
nest. This possibility is vividly coloured by the fact that a queen
ant, having mated for the only time in her life and then having
flown to a suitable place to found her new underground nest,
loses her wings. In some species she bites them off, in others she
rips them off with her legs. To bite off your own wings is pretty

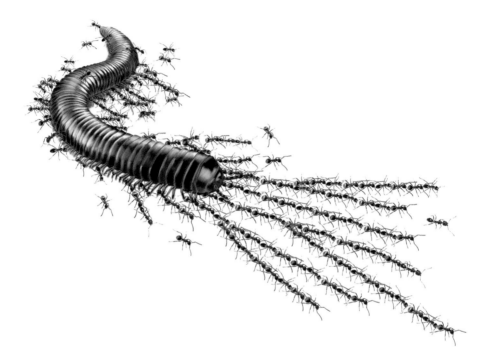

ANT CHAIN GANG

Ants are the great cooperators. In this case they form long lines to
drag a millipede which is far too big for any one ant on her own.

drastic testimony that wings aren't always desirable. They've served their purpose on the mating flight and the search for a new nest site. Surplus to requirements and probably an active hindrance underground, they are thrown away. Or eaten.

Admittedly worker ants don't spend all their time underground. They scuttle about foraging for food which they bring back to the nest. Even if wings are a handicap underground, mightn't it still be a good idea to keep them so the workers could forage fast like their wasp ancestors? Well, wasps may be faster than ants but consider this. Foraging ants often drag home to their nest great lumps of food heavier than themselves: a whole beetle, for instance. They couldn't fly with such a burden. Often they collaborate in teams to drag even larger prey. Teams of army ants have even been seen dragging a whole scorpion along. Where wasps and bees forage over large distances for small parcels of food, ants specialise in food that is relatively close to home and which can be too large to carry in flight. Even without a full cargo, flying is very energy-intensive. As we'll see later, wasp flight muscles are little reciprocating engines, and they burn a lot of sugary aviation fuel. Wings themselves must cost something to grow. Any limb has to be made of materials that enter the body as food, and four wings for every one of the thousands of workers in a nest would not be cheap to grow. They'd be a heavy drain on the colony's economic resources. Probably all these considerations tipped the workers' balance towards not growing wings. 'Tipped the balance' is an apt phrase, and we'll continue to meet the idea of economic balance again throughout this book. Questions of evolutionary advantage – what is this organ good for? – always involve an economic calculation of trade-offs – balancing the benefits against the costs.

Termites are very different from ants in some ways, not in others. When I was a child in Africa we called them 'white ants' but they aren't ants, not even close. Where ants are related to

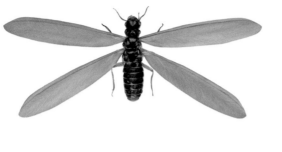

TERMITE QUEEN

ONCE HAD WINGS

Now she is reduced to nothing more
than a giga-factory for eggs.
The grotesque swelling of her abdomen
has stretched the brown exoskeletal
plates apart.

wasps and bees, termites are closer to cockroaches. In their evolution, they independently converged towards an ant-like way of life from their cockroach-like beginnings, as ants evolved from their wasp-like beginnings. But there are important differences between the two outcomes. Where worker ants, bees and wasps are always sterile females, worker termites are sterile males as well as sterile females. But they are like ants in that the workers are wingless while the reproductive females and males (queens and kings) have wings, which they use for the same purpose as winged ants. And winged termites swarm in a similar way to ants – rather spectacularly at certain times of the year. I had childhood friends in Africa who, when the winged 'white ants' were swarming, used to rush about stuffing them into their mouths – and, toasted, they were a local delicacy. As with ants, and presumably for the same reasons (termites typically spend even more time in enclosed spaces than ants), queen termites shed their wings after the mating flight. Indeed, they turn into grotesquely swollen shapes, for whom the very idea of wings would seem like a joke. The head, thorax and legs are unmistakably those of an insect, but the abdomen is a massively bloated, fat, white bag of eggs. The queen is just a walking egg-factory – actually not even a walking one, as she is too fat to walk. She'll churn out more than 100 million eggs during her long life.

Worker ants and termites are a telling example to begin this chapter, because every one of them is genetically equipped to grow wings but refrains from doing so. Queen ants, as we've seen, even rip or bite their wings off. No birds bite their wings off. It's hard to even imagine. The only remotely similar example I can think of among vertebrates is *autotomy* of the tail. From the Greek for self-cutting, autotomy is the shedding of the tail, or part of it, when a predator has caught it. It's a useful trick which has arisen many times independently in lizards and amphibians. But never in birds. Unlike queen ants, no bird autotomises its wings. Over evolutionary time, however, plenty of birds have gradually shrunk their wings, or even lost them altogether. Especially on islands – where more than sixty species of birds today (many more if you count extinct species) are known to have become flightless: among them geese, ducks, parrots, falcons, cranes and more than thirty species of rail, including the tiny Inaccessible Island rail of Tristan da Cunha.

Why do island birds lose the power of flight over evolutionary time? As we saw in the previous chapter, flightless birds are often found on islands too remote to have been reached by mammal predators or competitors. The lack of mammals has two effects. Firstly, birds, having arrived on wings, are able to take over the ways of life that would normally be filled by mammals; ways of life that don't require wings. The role of large mammals in New Zealand was filled by the now extinct flightless moas. Kiwis behave like medium-sized mammals. And the role of small mammals in New Zealand is (or was) filled by a flightless wren, the St Stephens Island wren (recently extinct), and by flightless insects, giant crickets called wetas. All are descended from winged ancestors.

Secondly, given that there are no mammal predators on their island, birds 'discover' that wings aren't necessary to escape being eaten. This is, presumably, the story for the dodos of Mauritius, and related flightless birds on neighbouring islands, descended from flying pigeons of some kind.

I put 'discover' in quotation marks for a reason. Obviously, those ancestral pigeons, newly touched down in Mauritius or Rodriguez, didn't look around and say, 'Oh goody, no predators, let's all shrink our wings.' What really happened over many

generations is that those individuals who happened to have genes for slightly smaller wings than average were more successful. Probably because they saved on the economic costs of growing them. They, therefore, could afford to rear more children, who inherited the slightly reduced wings. And so, as the generations went by, the wings steadily shrank. At the same time the bodies of the pigeons got larger. You could see this as diverting to other parts of the body resources saved through not needing to grow and service wings. Flying consumes plenty of energy, and diverting all that energy into other things, including increased size, makes a lot of sense. But getting larger in evolution is a general feature of island animals, so there may be more to it than that. Confusingly, in some cases, island species tend to get smaller. As we'll see in the next chapter, it's been suggested that species that arrive on the island large tend to get smaller, while those who arrive small get larger.

Bats, for obvious reasons, are often the only mammals capable of colonising remote islands. Yet I know of no examples of bats losing the power of flight, whether on islands or anywhere else. I find this surprising. You'd think the same reasoning that lies behind multiple evolutions of flightless birds on islands would apply to bats. I wonder whether, just possibly, they simply haven't been noticed. Maybe future molecular genetics will discover an island species of 'shrew' which will turn out to have sprung (in the evolutionary sense) from within the bats. It's fun to make speculations like that. If we seem to be wrong so far, there's always a chance that later research will prove us right. Stranger things have happened. Who would have guessed, before molecular genetics came along, that whales spring from right in the middle of the cloven-hoofed animals? Hippos are closer cousins to whales than they are to pigs! Whales are cloven-hoofed animals even though they no longer have hooves to cleave!

Dodos may have lost their wings because of a lack of predators. But unfortunately the poor dodos didn't survive the advent of sailors in the seventeenth century. It has been suggested that 'dodo' comes from a Portuguese word meaning 'fool'. They were fools because they didn't run away from the sailors who clubbed them for 'sport'. But the reason they didn't run away is presumably that the island had hitherto contained nothing worth running away from; the same reason their ancestors lost their wings in the first place. Probably a more important cause of their extinction than being clubbed for 'sport', or hunted for meat (contemporary reports suggested they didn't taste nice), was the rats, pigs and religious refugees that arrived in ships and competed with the dodos for food and ate their eggs.

Galapagos flightless cormorants are obviously descended from cormorants that flew to the islands from the mainland, and their descendants then lost their wings. All cormorants have a habit of hanging out their wings to dry after diving for fish. This is important because, when they dive for fish, their wings get sodden and are no good for flying. The same is not true of most water birds, who oil their feathers – hence the expression, 'Like

HANGING THEM OUT TO DRY

*The ancestors of the Galapagos flightless cormorant flew to the archipelago
on wings as big and well feathered as those of the mainland cormorant.
Once there, their wings shrank over evolutionary time. But Galapagos
cormorants still retain the ancestral habit of hanging them out to dry.*

DID TERROR BIRDS SWALLOW PREY WHOLE?

The cowering capybara is in danger of ending up inside the towering terror bird. To give an idea of scale, capybaras are giant guinea pigs, the size of a sheep. Terror birds are extinct (you may be happy to hear). Capybaras are still with us (you may be equally happy to hear).

water off a duck's back'. Galapagos cormorants still hang their wings out to dry although they can't fly even with dry wings. I should add that not all ornithologists buy the theory that drying them to prepare for flight is the only reason cormorants hang their wings out.

Dodos and Galapagos cormorants lost their wings relatively recently, within the last few million years. Ostriches and their kind lost their wings much earlier, presumably on long-forgotten islands to which their remote ancestors flew on fully developed wings. The wings which once carried their ancestors aloft have shrunk to short, stubby remnants. Or, in the case of the (extinct) moas of New Zealand, disappeared altogether. Ostriches use what's left of their wings partly for display to other ostriches, and partly to help them steer and balance as they run. This is especially necessary when running fast, and ostriches do run very fast.

There's also a suggestion that ostrich wings may be deployed when the bird needs to slow down – just as some aircraft release a parachute behind them when landing on an icy or short runway. Rheas, South American relatives of ostriches (which Darwin actually called ostriches), have somewhat larger wings in proportion, but still nowhere near big enough to fly. Rheas and ostriches are related also to the emus of Australia and the extinct moas of New Zealand: they're all *ratites*, and so are kiwis.

The 'terror birds' and their relatives, who finally went extinct only a couple of million years ago in South America were not ratites. Unlike ratites they were voracious carnivores, and their formidable name was well deserved. The largest of the terror birds stood 3 metres tall. Ratites are mostly vegetarian, with small heads and thin necks. The terror birds, of which there were many species, had huge heads and capacious necks. I can't help wondering whether they swallowed large prey whole, as other birds can do. Maybe even a capybara – a kind of giant guinea pig. And lest 'guinea pig' should mislead you into misjudging its size and hence that of the terror bird, I hasten to explain that an adult capybara can be a metre long. We're talking about a guinea pig as big as a well-grown sheep. Gulls have been seen swallowing rabbits whole, and also chicks from neighbouring nests in the gull colony. South America was home to even bigger giant guinea pigs, as big as hippos. They are now extinct and, although they were contemporaries of some terror birds, they were presumably too large to be threatened by them – at least not swallowed whole! But a capybara the size of a sheep? From a terror bird's height, wouldn't it look a bit like a rabbit does to a gull?

The shoebill stork, a magnificently ugly endangered species from Africa, is not a close relative of the terror birds, and it is small enough to be (just) capable of flying. But its appearance – and its feeding habits – give us an idea of what it might feel like to be just about to be swallowed whole.

The giant moas of New Zealand reached the same height as the large terror birds, much bigger than ostriches. Where the

IMAGINE CONFRONTING THIS
IF IT WERE 3 METRES TALL
The shoebill stork is too small to swallow
you. But its baleful stare gives an idea
of what it might have been like to be
confronted by a terror bird.

wings of most ratites (and terror birds) are small, moas went further and lost their wings altogether. Even whales haven't gone so far in the direction of limb loss. They've lost their hind legs

THE ROC OF THE ARABIAN NIGHTS

The elephant-carrying Roc never existed, nor could it. But did the legend arise from travellers' tales about the giant flightless elephant bird of Madagascar?

but they still have traces of leg bones left inside them. Moa wing bones are totally absent. Tragically, they were driven extinct by the newly arrived Maoris. This happened only about 600 years ago, but even so my New Zealand friend was surely mistaken when he regaled me in a pub with stories about hearing them bellowing to each other in the South Island bush.

The Maoris arrived in New Zealand about 700 years ago – just yesterday compared with the 50,000-odd years since Aboriginals reached Australia. It's controversial whether Aboriginals were responsible for the extinction of the many large marsupial mammals that used to live in Australia. There were also huge flightless birds such as *Genyornis*, 2 metres high, a kind of overgrown goose. These Australian 'thunder birds' were not closely related to ratites – nor to terror birds, whose closest living relatives are the seriemas of South America, elegantly crested, long-legged birds who stand but a fraction the height of a terror bird.

Also gigantic were the so-called elephant birds of Madagascar, flightless ratites again. There were several species of elephant bird. The largest, recently renamed *Vorombe titan*, stood 3 metres tall. And now for a tantalising flight of fancy. The story of Sinbad the Sailor is one of the colourful legends of *The Arabian Nights*. Among Sinbad's more alarming adventures was his encounter on an island with a giant bird called the Roc, which fed its young on elephants. Sinbad needed a lift off the island, so hitched himself by means of his unravelled turban to the Roc's mighty

talons while it was sitting on its equally mighty egg.

Marco Polo, the medieval Venetian explorer, also mentions the Roc. He said it was so huge it would seize elephants and drop them from a great height to kill them. Interestingly, it seems he believed the Roc came from Madagascar. Madagascar? That's where we find the remains of the elephant birds. Perhaps the legend of the Roc started from travellers' tales about gigantic birds in Madagascar, successive rumours exaggerating their size and forgetting the important fact – known to eye-witnesses but not to rumour-spreaders – that they couldn't fly. Elephant birds went extinct only recently, perhaps as late as the fourteenth century; probably, like the moas, exterminated by newly arrived humans eating them and their eggs, clearing the forest for agriculture and destroying the habitat of the great birds. There seems to be some hope that they could one day be revived, perhaps from DNA extracted from their eggshells, which are still to be found in abundance on Madagascar beaches. Maybe moas too. Wouldn't that be wonderful? Incidentally, it's a surprising fact that the closest living relative of the giant elephant birds is the smallest of all the ratites, the kiwi of New Zealand.

On a Madagascar beach, David Attenborough paid people to find fragments, which he and a film-crew colleague stuck together with sticky tape, to reconstruct the almost complete shell of an elephant bird egg. Such eggs had a volume about 150 times that of your breakfast egg. That's breakfast for a whole company of soldiers. Elephant bird eggshells are remarkably thick, about as thick as the reinforced glass of a car's windscreen. You'd think you'd need an axe to open the egg when making the soldiers' breakfast. That makes us wonder how the chicks got out.

> ‭ **By the way**, this is another case of evolution being full of trade-offs – compromise – as in human economics. Where eggshells are concerned, the thicker they are, the safer they are from breakage by predators or by the weight of the parent sitting on them. But, by the same token, thick shells are hard for the chick to crack when the moment comes to hatch. And the thicker they are, the more costly they are in precious resources such as calcium. Evolutionary theorists love to speak of trade-offs between 'selection pressures'. Different selection pressures continuously nudge the evolving species in different directions, resulting in a multiway compromise. Natural selection from predators exerts a pressure, in

evolutionary time, to evolve thick shells. But at the same time there is an opposing pressure towards thinner shells, as some baby chicks die trapped inside eggs with thick, tough shells. Those babies that are least likely to be trapped inherit the genes for making thinner eggshells. On the other hand, those same genes make shells that predators can break into easily. Some chicks die for one reason while other chicks die for the opposite reason, where eggshell thickness is concerned. As the generations go by, the average shell settles down to an intermediate thickness, a compromise between the opposing pressures.

In birds that fly, another pressure comes from the need to be light. Flying birds go to great lengths to reduce their weight, with hollow bones and nine air sacs in various parts of the body. Much of the good achieved by these measures would be undone by heavy eggs. That is no doubt why birds carry only one fully formed egg inside them at a time. A clutch may consist of many eggs, but the parent doesn't start incubating them until the last one is laid, so that the chicks hatch at the same time. Some birds of prey show us an additional, rather cruel compromise. Mothers lay more eggs than they expect to rear. If it's an exceptionally good food year they may rear all of them. But in a normal year the smallest chick is expected to die, often murdered by its siblings. The smallest chick can also be seen as an insurance premium on the life of the older ones.

Mammals, typically, are different. Lacking the same selection pressure to be ultra-light, pregnant mammals often carry many embryos simultaneously (the record, thirty-two, is held by a Madagascan tenrec; it looks a bit like a spiky hedgehog and you can't help feeling for the mother giving birth). But not bats, whose litter size is typically one, for the same reason as we just saw for the birds. Not humans either, but that's for a different reason. We don't have big litters probably because of our big brains. Whatever the (no doubt good) reasons for our big brains, they make childbirth exceptionally difficult and painful. Before modern medicine, a shockingly high percentage of women died in childbirth, and the main problem was the baby's enormous head. Once again we see compromise in evolution. Human babies reduce the danger to their mothers by being born relatively early in development, but not so early as to endanger their own survival. Their heads are still too large for the mother's comfort; and twins or larger litters increase the problem. Forced to be born early, human babies are unusually dependent on their parents compared with other large mammals. We can't walk till we're about a year old. Baby gnus (wildebeest) can walk the day they are born. They too are born singly because they have to be capable of walking – running, even – pretty much straight out of the womb. If they were born in a large litter, they wouldn't be big enough to keep up with the migrating herd.

Human technology is rife with pressures pushing in incompatible directions. In this case the pressures operate not in evolutionary time but on the timescale of successive designs on drawing boards. Planes need to be as light as possible, like birds. But, like eggshells, they also need to be strong. The two ideals are incompatible, so a compromise has to be struck – balance, as we saw before. Air travel could be safer than it is. But at a cost not only in money but in inconvenient nuisance and delays. Yet again, a balance has to be struck. If safety were infinitely valued, every passenger could be strip-searched and every suitcase turned inside out by security guards. But the acceptable trade-off stops short of such draconian extremes. We accept some risk. Disagreeable as the thought might be to starry-eyed idealists not used to thinking like economists, human life is not infinitely precious. We put monetary value on it. The regulations for military and civilian aircraft are stabilised at different safety compromises. Economic trade-offs, balance and compromise are fundamental to both technology and evolution and these ideas crop up throughout this book.

Why are bats the only mammals that truly fly? Actually, bats represent a respectable proportion of all mammals. About a fifth of all mammal species are bats. But why don't we see winged lions roaring through the air after winged antelopes? As it happens, that's actually an easy one to answer. Lions and antelopes are too big. But flying rats? About 40 percent of all mammal species are rodents. Why have none of them developed wings while they've scuttled and whiskered and gnawed their way through 50 million years of evolutionary history? Perhaps the answer is that bats got there first. If some viral pandemic were to exterminate all bats, my guess is that rodents would rise to the occasion, not just as gliders (they already do that) but as true flyers. But we mustn't forget economics. Wings are costly to grow and even more costly to use, especially to flap. They have to justify their costs. And, as we saw with the ants, wings can get in the way. If you make your living underground as a naked mole-rat (delightfully ugly little burrowers who live in social groups with a super-reproducing 'queen', a bit like ants or termites) does, wings would be a positive handicap.

We now begin our list of ways in which animals have managed to get off the ground in the face of gravity. The easiest, least laborious way to get off the ground is perhaps the simplest: go to the opposite extreme from the mythical Roc or the real ostrich or terror bird. Don't be large. Be small.

CHAPTER 4

FLYING IS
EASY IF YOU
ARE SMALL

CHAPTER 4

FLYING IS EASY IF YOU ARE SMALL

It's a pity the Cottingley fairies didn't exist because, unlike angels or Buraq or Pegasus, those imaginary little people were the right size to find flying easy. Flying becomes more and more difficult the bigger you are. If you are as small as a pollen grain or a gnat you hardly need the effort of flying. You can pretty much just float in the breeze. But if you are as big as a horse, flying becomes an immense effort, if not downright impossible. Why does size matter? The reason is interesting. Now we need a little mathematics.

If you double the size of anything (say its length, with all other dimensions growing in proportion), you might think its volume and weight would double too. But it actually gets eight times heavier ($2 \times 2 \times 2$). This is true of any shape that you might scale up, including people, birds, bats, planes, insects and horses, but we can see it most clearly with square children's bricks. Take one cubical brick. Now stack bricks to make the same shape but twice as big. How many bricks are there in the

SMALL THINGS HAVE A RELATIVELY LARGE
SURFACE AREA
*If you scale something up, its volume (and hence weight) increases
more steeply than its surface area. This is most easily seen with
bricks, but it's true of everything, including animals.*

larger stack? Eight. The double-sized block of bricks weighs eight times as much as the single brick of the same shape. If you now make a stack three times as big, you'll find you need twenty-seven bricks: 3 × 3 × 3, or 3-cubed. And if you try to build a stack measuring ten bricks in each direction you'll probably run out of bricks because 10-cubed (1,000) is an awful lot of bricks.

Take any shape and multiply its size by some number to scale it up. The volume (and weight, which obviously affects flying) of the inflated object will always go up as the cube of the number: the number multiplied by itself twice. The calculation applies not just to bricks but to any shape you care to scale. But while the weight of an object as you grow rises as the cube, the surface area only goes up as the square. Measure the amount of paint you'd need to cover one brick. Now scale the stack up so that it measures two bricks in every direction. How much paint do you need to cover it? Not twice as much and not eight times as much. You'd need only four times as much paint. Scale the stack up ten times, so it measures ten bricks in every direction. We've already seen that it would weigh 1,000 times as much: you'd need 1,000 times as much wood. But only 100 times as much paint. So the smaller you are, the larger your surface area relative to your weight. We'll return to surface area and why it matters, in the next chapter. Here, it's enough to say that a large surface catches the air.

Following the flights of fancy with which we began, let's think of an angel as a person with wings, a scaled-up fairy. The archangel Gabriel is typically portrayed in paintings as about the same height as a normal human, say 170 centimetres. About ten times as tall as a Cottingley fairy. So Gabriel would be not ten times heavier than the fairy but 1000 times heavier. Think how much harder the wings would have to work in order to lift the angel. And the scaled-up wings wouldn't have a thousand times the area but only a hundred.

DID LEONARDO WONDER WHETHER GABRIEL'S
WINGS WERE TOO SMALL?
The Annunciation – *but with wings made big enough to get Gabriel*
off the ground. Even so, where would he keep the massive breast muscles
necessary to power them? And the breastbone 'keel' to attach the
muscles? Leonardo was too good an anatomist not to have wondered.

If you've visited the Uffizi gallery in Florence, you will have
seen Leonardo da Vinci's ravishingly beautiful *Annunciation*.
It shows the angel Gabriel but with surprisingly small wings.
Those wings would struggle to lift a child, let alone the man-
sized (albeit woman-featured) Gabriel that Leonardo painted.
And it has been suggested that Leonardo painted the wings
even smaller at first, and they were enlarged by a later artist.
Not enlarged enough. Not nearly enough. We've doctored our

TINY HUMMINGBIRD, HUGE KEEL
See how relatively big the breastbone 'keel'
is, even in such a tiny bird. It has to be big, to
support the expensive flight muscles.

reproduction of the painting to make the wings a bit more fit
for purpose. Unfortunately, it spoils the beauty of the picture.
That's putting it mildly. They reach absurdly outside the frame.

In Leonardo's *Annunciation*, the root of the wings – unlike the
rest of his exquisite picture – is so awkwardly drawn, it's almost

as though he was embarrassed by the absurdity of it. Where, the great anatomist perhaps wondered, do angels keep the massive flight muscles they'd need. And the breastbone to attach them? If he'd painted the necessary keel it would have reached a fair way towards the table where the Virgin Mary is sitting. Pegasus, being a horse and much heavier, would have needed an even deeper keel. Buraq's keel would bump along the ground when the poor creature tried to walk. Look at the relatively huge keel of the hummingbird, one of the smallest of birds but a very vigorous flyer. Think how relatively much bigger Pegasus's keel would have to be. Bats, actually, don't have the same kind of keel as birds, but other breastbones are enlarged and strengthened to do the same job.

It is certain that the wings on Leonardo's Gabriel are too small. But how might we go about calculating the actual size of wings that a human-sized creature would need, in order to fly? It would be simpler if, like Boeing or Airbus designers, we could use the mathematics of fixed-wing aircraft. And that's difficult enough. But living wings adjust their shape from moment to moment. To make matters worse, they flap in complicated patterns, and the consequent eddies and swirls of air make the calculations even harder. Probably our easiest course is to give up on theoretical calculations and look around the world for a bird as big as a human.

All the largest birds today are flightless, like ostriches. But there are a few extinct ones that flew, and whose weight was in the same ballpark as a person's. *Pelagornis* was a gigantic seabird. It probably lived and flew like an albatross, and had the same slender wings but twice as long. Unlike albatrosses it had teeth – well, not real teeth but spikes along the beak that look like teeth and would have worked like teeth, probably helping to snag fish and stop them escaping. We'll see later that albatrosses get most of their lift in a special, cunning way, exploiting the winds that shear the waves, and *Pelagornis* may well have done the same. It had a wingspan of about 6 metres.

Even bigger than *Pelagornis*, or at least heavier though with about the same wingspan, was *Argentavis magnificens*, whose Latin name translates appropriately as 'magnificent bird of Argentina'. *Argentavis* was probably related to today's Andean condor, itself a splendidly large bird (in danger of extinction, alas), but *Argentavis* was far bigger. It weighed some 80 kilograms, about the same as a well-built man, with the reservation that much of its weight must have lain in its wings themselves. Its wings were much less slender than albatross or *Pelagornis* wings, more square like a condor's. And much larger in area, as they'd need to be, to carry a bird which could outweigh ten albatrosses. The estimated wing area of *Argentavis* was about 8 square metres, about the same as a modern sport parachute. It's reasonable to think that *Argentavis* mostly glided and soared in up-currents like modern condors and vultures, who only occasionally flap.

Probably the largest flying animal ever was *Quetzalcoatlus*, not a bird but a pterosaur. Pterosaurs were a large group of

THE LARGEST BIRDS EVER TO FLY

The extinct Pelagornis *and* Argentavis *with parachutist for scale.*

QUETZALCOATLUS WAS PROBABLY
THE LARGEST ANIMAL THAT EVER FLEW
Of course, it never met a giraffe – they are separated by perhaps
70 million years. But if they could meet they'd eye each other head to
head. Can you imagine a giraffe taking to the air?

flying reptiles, commonly called 'pterodactyls', although that is technically the name of a particular kind of pterosaur, much smaller than *Quetzalcoatlus*. Strictly speaking pterosaurs were not true dinosaurs, but they were related to them and they finally disappeared at the same time as the dinosaurs in the great extinction at the end of the Cretaceous period.

Quetzalcoatlus was monstrously big. Its wingspan was 10 to 11 metres, comparable to a Piper Cub or Cessna aircraft, and bigger than any bird including *Argentavis*. If it stood up it could see eye to eye with a giraffe. And it probably did stand up, on its hind feet and its front knuckles with the wings folded. Thanks to its hollow bones (shared by all flying vertebrates) *Quetzalcoatlus* was only a quarter the weight of a giraffe, however. Like very large birds it probably spent most if not all of its air time gliding. Once airborne it could probably stay aloft for long periods and travel enormous distances at high speeds. *Quetzalcoatlus* pushed to the utmost limit the size at which flying with muscles is possible. I'd guess it preferred to glide from high places, but if it ever needed to take off from the ground that must have posed quite a problem. It may have used its powerful front limbs to 'pole-vault' into the air. How, you might wonder, could such a long neck support such a huge head in a flying animal? Recent research reveals that the bones of the neck vertebrae were mainly hollow (for lightness) with a network of stengthening struts radiating out, like the spokes of a bicycle wheel, from a hub through which ran the spinal nerve cord.

We don't know whether these gigantic, leathery old aeronauts could flap their wings or whether they only soared and glided. That's an important difference and we'll come back to it in later chapters.

⤳ **By the way**, flying isn't the only thing that gets harder when you get bigger. Walking does too. Even just standing up. Fairy-tale giants are pictured as normally shaped, if ugly, men but scaled up. If the bones of a 30-foot-tall ogre were like normal human bones but simply scaled up in proportion, they'd break under the weight. He'd weigh not just five times as much as even a tall 6-foot man but 125 times as much. To avoid collapsing in a heap of painful fractures, the giant's bones would have to be out-of-proportion thicker than normal human bones. Like the bones of elephants and large dinosaurs they'd be thick like tree trunks, so thick they'd look out of proportion to their length.

Size is one of the easiest things for animals to change in evolution, in either direction. As we saw when talking about dodos on Mauritius, animals that move to an island often evolve to become larger – *island gigantism*. Confusingly, island arrivals may under other circumstances evolve to become smaller – *island dwarfism* – like the 1-metre miniature elephants that used to live on Crete, Sicily and Malta, and which must have been charming. Foster's Rule states that previously small animals tend to get larger when they arrive on an island, while previously large animals get smaller. I'm not sure how clearly we understand why this is. A suggested reason is that hunted animals (which tend to be small) get larger because of the absence of predators. But large animals get smaller because the small area of islands limits their available food.

By now you will have worked out that an evolutionary change in size cannot be a simple scaling up or scaling down. The proportions must change too, because of those mathematical laws we met earlier with the toy wood bricks. The whole shape of the animal has to change. Animals that evolve smaller size become more spindly and spidery. Animals that evolve larger size have to develop thicker, more tree-trunky limbs. All the proportions of the animal have to change to go along with changes in absolute size, not just of bones but – as we'll see in the next chapter – heart, liver, lungs, intestines and other organs too. And all for mathematical reasons of the kind we met at the beginning of this chapter.

To return to the chapter's title, if you are really small like a fairy or a gnat, flying is easy. Like gossamer, like thistledown, the slightest puff of wind can carry you off. If you need wings

at all, they might be less for getting off the ground and more for steering.

The Cottingley fairies' wings could afford to be quite small, and they could be flapped without much muscular effort. The fairy in *Peter Pan* is called Tinkerbell. Charmingly, the smallest flying insect is the Fairyfly (it's actually a wasp but never mind) and the Latin name of one species of fairyfly is *Tinkerbella nana* ('nana' comes from the sheepdog nanny of the Darling children in *Peter Pan*). The gossamer 'feathers' of *Tinkerbella nana* are technically wings but it probably uses them more like oars to 'row' through the air in which it floats, rather than to provide any significant amount of lift. Other species of fairyfly have wings that look more like conventional wings. Fairyflies are the smallest flying animals so far known. Insects as small as this would have no problem staying aloft. On the contrary, it might be hard work coming down to earth.

Being small is all very well. But what if you need to be large for some reason and still need to fly? There are many good reasons to be large, even though the economic costs are high. Small animals are vulnerable to being eaten. They can't catch large prey. Rivals of your own species, rivals for mates perhaps, are easier to intimidate if you are bigger than them. If, for whatever reason, you can't be small and yet still need to fly, you have to find another solution to getting off the ground. This brings us to the next chapter.

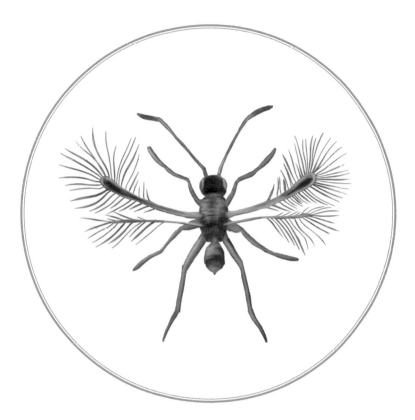

TINKERBELLA

The picture that heads this chapter shows it
flying through the eye of a needle.
Wingspan approximately 0.25 millimetres.

IF YOU MUST BE LARGE AND FLY, INCREASE YOUR SURFACE AREA OUT OF PROPORTION

FLYING SQUIRREL

Gliding squirrel, or parachuting squirrel, would be a better name.
The 'patagium', a flap of skin stretched between the limbs, increases the
animal's surface area and allows it to glide safely from tree to tree.

CHAPTER 5

IF YOU MUST BE LARGE AND FLY, INCREASE YOUR SURFACE AREA OUT OF PROPORTION

The previous chapter showed that small animals automatically have a relatively large surface area compared to their weight. This is why flying comes easily to them. We saw this in our little mathematical calculation involving children's bricks. We measured the surface area of a thing as the quantity of paint you'd need to cover it. Or the amount of cloth you'd need to clothe it. If an angel is the same shape as a fairy but ten times as tall, the area of skin covering the angel would be 10 squared or 100 times as great, while the volume and weight would be 1,000 times as great.

But what does surface area have to do with flying? The larger your surface, the larger the area available to catch the air. Take two identical balloons. Blow one of them up so it has a large surface area, and leave the other one as a little flabby bit of rubber. Drop them simultaneously from the Leaning Tower of

A BIPLANE

Slow planes need a relatively large wing area to support a
given weight. Biplanes are now much rarer than they used
to be, and there are no very fast biplanes.

Pisa. Which one will hit the ground first? The un-blown-up one
will, even though it is no heavier (actually it's slightly lighter). Of
course, if you dropped them in a vacuum they'd hit the ground
simultaneously (to be realistic, in a vacuum the inflated balloon
would explode, but you get the point). I said 'of course', but
that would have surprised everybody until Galileo came along.
He showed that even a feather and a cannonball would hit the
ground simultaneously if dropped in a vacuum.

Our question in this chapter is, what if an animal needs to be
large for some reason yet still needs to fly? It must compensate
by increasing its surface area disproportionately: grow

projections like feathers (if a bird) or flaps of thin skin (if a bat or a pterodactyl). However much material your body is made of (your volume or weight), if you splay some of that volume out into a big surface you will have taken a step towards flying. Or at least parachuting gently down, or floating in the breeze. This is why our made-up version of Leonardo's angel needed such colossal wings. Engineers express the point mathematically in terms of what they call wing loading. The wing loading of an aircraft is its weight divided by its wing area. The larger the wing loading, the harder it is to stay aloft.

The faster a plane – or a bird – flies, the more lift can be squeezed out of each square centimetre of wing. Faster planes of a given weight can tolerate a smaller wing area and still stay aloft. This explains why slow planes tend to have relatively larger wing areas than fast planes. Before today's high speeds were achieved, early planes were often biplanes. You get double the wing area, but also increased drag. For the same reason, triplanes were sometimes seen.

> **By the way**, leaving the subject of flight for a moment, the relationship between surface area and volume is very important in living bodies generally and it provides an interesting digression. Just as wings increase the external surface area, and this is important for flying, there are many organs that increase the internal surface area to keep pace with increasing body size. Lungs, for instance.

The volume or weight of an animal is a good measure of the number of its cells. Bigger animals don't have bigger cells, they just have more of them. Every one of those cells, whether in an elephant or a mouse, needs to be supplied with oxygen and other vital substances. A flea has fewer cells than an elephant and those cells are never far from the air. The oxygen hasn't got far to go to reach the cells. An adult person has about 30 trillion cells, and only a minute fraction of them are skin cells in contact with the air. Even though the surface area of the person is much larger than that of a flea, a smaller proportion of our cells are on our outer surface. We large animals compensate for our lack of external surface by growing a huge internal surface which is exposed to the air. That's what lungs are all about. Inside your lungs is an intricate system of branching and sub-branching tubes culminating in tiny chambers called *alveoli*. You have about 500 million alveoli and their total area, if spread out, would cover a good part of a tennis court. That entire surface inside you is exposed to the air and is richly supplied with blood vessels. Even insects, though much smaller, increase their surface exposed to the air with a system of internally branching air tubes, *tracheae*. It's as though the whole insect body is a lung.

Blood vessels in our lungs branch and branch and branch to provide a huge internal surface area for collecting oxygen from the lungs and distributing it to all the cells of the body, for example muscle cells where it is necessary for the slow combustion that powers them. Capillary blood vessels constitute a gigantic internal area for collection and distribution purposes, supplying all the cells. To survive, a typical cell needs to be within about 5 percent of a millimetre of the nearest capillary. That is to say, within about two or three cell diameters of the nearest capillary. Capillaries collect food substances from the intestines, which themselves provide a very large internal surface area, again the best part of a tennis court. Think of the huge length of coiled-up intestine inside you, and compare it with that of an earthworm – just a straight tube from one end of the worm to the other. Your kidneys are furnished with countless tiny tubes – which, again, add up to a huge internal surface area – where the blood is filtered and stripped of waste substances. If you stretched out all your blood vessels, most of them tiny capillaries, they'd reach right round the world more than three times. That represents a colossal surface area of contact between blood and cells. Many of the large organs inside your body, not just lungs and intestines but liver, kidneys, etc., are all about increasing the effective surface area of blood's access to cells. As it happens, too, the crannies and crevices of a coral reef, the rutted tree bark and innumerable leaves of a forest, hugely magnify the surface area available for life to transact its business.

The conclusion from the digression is that the title of this chapter, 'If you must be large, increase your surface area', applies not just to flying but to breathing, blood circulation, digestion, waste disposal and just about everything that goes on inside an animal, as well as what we can see on the outside. But now let's return to flight.

As we realised previously, the larger the surface area of an animal compared to its weight, the slower it will fall through the air, and the more easily it will obtain the lift necessary to fly. Wings, whether used for flapping or gliding, are the obvious surface-area promoters. In bats and pterosaurs they are thin sheets of skin. Thin surfaces need support – bony support or something equivalent. Evolution is opportunistic: it tends to modify what is already there rather than sprout something completely new. In theory you could imagine wings springing from the back as in pictures of angels. But that would mean growing new supporting bones. What is available by way of existing bones which might be commandeered to provide support for flight surfaces? As we'll see later, there are lizards that glide on a skinny membrane that sticks out sideways and which borrows the ribs for support. But more professional flyers like bats, birds and pterosaurs make use of the arms, which are already supplied with serviceable bones and muscles, ripe for modification.

Bats and pterosaurs stretch their flight skin between the arms and same-side legs. Pterosaurs kept most of the arm bones relatively short, but hugely grew one finger, the fourth or 'ring' finger. 'Pterodactyl' literally means 'wing finger'. Almost the entire support to the front of the wing is provided by this one enlarged finger which reaches right out to the tip of the wing. Our fingers are fine, delicate things, which we can use for skills like typing or playing the piano. It's hard for us to imagine a single finger growing longer than the rest of the arm put together, and strong enough to support a big wing like that of *Quetzalcoatlus*. It even makes me a little squeamish to think of it. It's a lesson in what evolution can do by way of exploiting what is already there. I should say that the wing membrane itself doesn't fossilise well, so reconstructions by different biologists don't always agree. We have followed the most authoritative recent reconstructions in drawing the wing as stretching all the way to the ankle. There is also some evidence of a tendon running along the hind edge of the wing from fingertip to ankle, providing additional support and presumably preventing juddering in the wind – which would not only impair flight efficiency but also perhaps tear the wing.

Bat wings make use of all the fingers, not just the fourth. And, like pterosaurs, bats use the hind leg as an extra supporting strut for the wing. This condemns them to be poor walkers. The best walkers among bats are probably the short-tailed bats which shuffle through the leaf litter of the New Zealand forests.

o 1

o 2

o 3

But they are no match for a bird when it comes to walking or running. I imagine a walking pterosaur flopping along like an animated broken umbrella.

Birds do things differently. Instead of a flap of skin, their flight surfaces are made of cunningly spreadable feathers. The feather is one of the wonders of the world, a marvellous device, strong enough to support the bird in the air but less rigid than bones. Flexible as they are, feathers are at the same time stiff enough to enable the bird wing to economise on bone. In some birds, such as the raven pictured, the bony arm skeleton reaches only about halfway along the length of the wing. The rest of the wing's span is provided by feathers. Compare it to a bat or a pterosaur, where bone reaches all the way to the wing's extremity. Bones are strong but they are heavy, and heavy is exactly what you don't want to be if you are a flyer. A hollow tube is much lighter than a solid rod and only slightly less strong. All flying vertebrates have bones that are as hollow as they can get away with, strengthened by cross-struts. Birds get away with as little bone in the wing as possible, making use of the stiffness of ultra-light feathers instead.

THREE WAYS TO TURN AN ARM INTO A WING
Bats (01) have elongated all their fingers and spread them.
Pterosaurs (02) hugely enlarged just one finger. Bats and pterosaurs
need to incorporate the leg for additional support. Birds (03) don't,
because feathers have a stiffness of their own. And their arm bones can be
surprisingly (and economically) short, for the same reason.

Robert Hooke, in his 1665 book *Micrographia*, was one of the first to make drawings looking through a microscope and he astonished his readers with the intricate, fine structure of living bodies. Not surprisingly, the feather caught his attention, 'For here we may observe Nature, as 'twere, put to its shifts, to make a substance, which shall be both light enough, and very stiff and strong.' He goes on to note that 'very strong bodies are for the most part very heavie also' and, therefore, that if the feather had been constructed in any other way than the way it actually was constructed, it would have been much heavier. Wing feathers slide over each other, so the wing behaves like a perfect fan as it changes shape to accommodate different flying conditions. In this respect the bird wing is superior to that of a bat or pterosaur, where the price to be paid for changing wing shape is loose folds of dangling skin. The vanes of a feather are made of hundreds of *barbules* which zip or unzip together with their neighbouring barbules. This arrangement achieves Hooke's ideal of strength with lightness, but at a cost: the feathers require constant attention from the beak to keep them well zipped and in good order. If you watch a bird for any length of time you are sure to see it preen, paying close attention to the wing. A bird's life literally depends on it, for badly zipped wing feathers could be the direct cause of poor

flight performance and hence failure to escape from
a predator. Or failure to catch prey. Or a failure of
steering leading to a collision.

Feathers are modified reptile scales. They probably
originally evolved not for flight but for heat insulation
like mammal hair. Once again we see evolution taking
advantage of what's already there. (Another example:
male desert sandgrouse fly many miles to fetch water
for their young. Their belly feathers are modified to act
as a sponge. They fly it home where their chicks suck the
water.) Fluffy, insulating feathers later became longer
with a strengthened supporting quill down the middle,
and their tough flexibility was perfect for flying. A bird's
wing is a whole flight surface made of feathers, and its
area is large compared to the rest of the bird's surface.
The so-called primary feathers do most of the work of
flying. They're the big feathers that typically provided
the quills that our ancestors sharpened as pens.

It has only recently been discovered that, before
true birds evolved, feathers were common among the
group of dinosaurs from which the birds sprang. It
even seems likely that the dreaded *Tyrannosaurus*
had feathers, which somehow makes it seem just
a little less terrifying – if not exactly cuddly.
And there were four-winged dinosaurs with
feathers. They lived in the Cretaceous

FOUR-WINGED DINOSAUR
*A design that birds might have
adopted but didn't.*

period, 120 million years ago, actually later than the famous *Archaeopteryx*, which is often credited as the first bird. It seems probable that creatures like the *Microraptor* (pictured) were capable of true flapping flight, not just gliding.

Feathers are stiff enough that the wings need no supporting bones behind the arm, and the arm skeleton itself can be economically much shorter than the wing. At the same time, the cunningly curved feathers are bendy enough to work well on the upstroke as well as the downstroke. Even better, there's no need for the hind limbs to be co-opted into stiffening the wings. This means that birds, unlike bats and pterosaurs, are excellent walkers, runners and (in the case of small birds) hoppers. This is a huge advantage compared with clumsily lolloping pterosaurs or bats.

Insects have the same advantage. Not involved in flying, all six limbs are free for walking and running. A tiger beetle can fly, for example when it needs to escape from a lizard, but it mainly hunts prey like spiders or ants on foot. And when hunting it can run at 2.5 metres per second. That's about 125 body lengths per second. It's not really fair to turn that speed into the human equivalent, but I can't stop you doing the calculation for fun if you want to. And just look at the tiger beetle's long, splendidly athletic legs.

TIGER BEETLE
Champion sprinter of the insect world;
yet it can still fly.

Insect wings have no specific struts supporting them unlike the bone-stiffened wings of flying vertebrates. The skeleton of an insect is in any case an *exoskeleton*. The whole outer body wall of an insect is its skeleton. Wings are outgrowths of the exoskeleton of the thorax and so are automatically stiff enough to bear the load of a small flying animal.

For this chapter, what matters is that wings have a large surface area compared to the size of the animal as a whole. Such a large surface area is needed to provide lift in air. The winged sandals of Hermes (Roman Mercury) the messenger of the Greek gods were much too small, as absurd as the dinky little propellers in this doomed, though charming, Victorian design for a flying machine.

WOULDN'T IT BE NICE IF FLYING WERE THIS EASY?
Victorian design, flying in doomed fancy over Matthew Arnold's
'Sweet City with her Dreaming Spires'. How appropriate for Oxford,
his 'Home of lost causes and forsaken beliefs'.

CHAPTER 6

UNPOWERED FLIGHT: PARACHUTING AND GLIDING

CHAPTER 6

UNPOWERED FLIGHT:
PARACHUTING AND GLIDING

No matter how heavy you are, if you have a large enough surface area you can tame gravity to the extent of floating down gently and safely. That's what we do with our parachutes. This chapter will look at parachuting – and gliding – using surface area extensions that may include wings, but we'll begin with surface extensions that are not really wings.

Very small animals, as we've seen, automatically have a large surface area relative to their weight, and they can float down safely without the need of a specially constructed parachute. Squirrels aren't quite small enough for that. They need just a little help from an increased surface area. Skilled, fast climbers, they speed up their progress by leaping from branch to neighbouring branch. Their long, feathery tail increases their surface area and this helps them leap to a branch which is a little further than they could safely reach without the feathery tail. It's not a true flight surface like a wing, but every little helps, and the squirrel

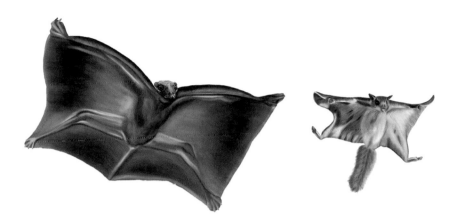

TWO INDEPENDENTLY EVOLVED LIVING PARACHUTES
Colugo or 'flying lemur' (left) and 'flying squirrel' (right).

is small enough to benefit from a bushy tail as an air-catching surface.

There are some specialist squirrels, so-called flying squirrels (they should better be called gliding squirrels), who have carried the idea further. They have evolved a web of skin stretching from front leg to hind leg, which amounts to a parachute. It's called the *patagium* (from the Latin for the edging on a Roman woman's tunic). Flying squirrels can not only jump from branch to branch: they stretch out their arms and legs to deploy their parachute, and they glide gently from one tree to another tree which might be 20 metres distant. As with our parachutes, they are drifting downwards, but the downward drift is slow and safe, and it takes them to a different tree in the forest. They typically glide from high in one tree to near the base of the trunk of another tree.

The forests of South East Asia and the Philippines house a creature that takes the idea a little further. The colugo, cobego or 'flying lemur' is not really a lemur (all true lemurs live in Madagascar). They aren't classified as primates (the group of mammals to which lemurs and monkeys and we belong) but colugos are related to primates. Like flying squirrels they have evolved a patagium. But it doesn't just stretch from arm to leg. It incorporates the tail too. Pretty much the whole body amounts to one big parachute. Their patagium has a bigger surface area than a flying squirrel's and they can glide 100 metres. Again, the patagium is not a true wing. It isn't flapped like the wing of a bat or a bird. But by adjusting its limbs the colugo can steer its glide, much as an expert human parachutist does by tugging on ropes. Actually, although most flying squirrels don't incorporate the tail in the patagium, there's a giant flying squirrel in China whose patagium stretches a short distance along the tail. This gives a clue as to how the colugo parachute might gradually have evolved.

Colugos and squirrels have evolved the patagium independently – convergent evolution. But they aren't the only forest-dwelling mammals to have done so. Australia has been isolated for most of the time since the dinosaurs went extinct and the mammals took over their dominant role on land. In Australia it happened that the mammals that were on the spot to step

into the dinosaurs' empty shoes were all marsupials (plus a few egg-laying mammals: ancestors of platypuses and spiny anteaters). A huge range of marsupials evolved in Australia and New Guinea, paralleling all the mammals familiar in the rest of the world. There were marsupial 'wolves', marsupial 'lions' and marsupial 'mice'. The quotation marks indicate that these were independently evolved 'wolves', 'lions' and 'mice', not what we in the rest of the world would call real ones. There were also marsupial 'moles', marsupial 'rabbits' and – you've guessed it – marsupial 'flying squirrels'. These Australian marsupial gliders are called flying phalangers. I should add that for many zoological purposes, including this one, the big neighbouring island of New Guinea counts as Australia. New Guinea has its own kangaroos among its own marsupial fauna. And it has its own marsupial gliders similar to the Australian ones.

There are several species of marsupial glider. They all resemble flying squirrels in that the patagium stretches from arm to leg, but doesn't incorporate the tail as in the colugo. The one that most resembles a flying squirrel is the sugar glider, which occurs in both Australia and New Guinea. It can glide to a distant tree about 50 metres away. It looks like a twin of a flying squirrel, but it's almost as distantly related as it is possible to be while still being a mammal. Such convergent evolution is a beautiful example of the power of natural selection. A patagium is a good thing for a forest-dwelling mammal to have. So it evolved independently in both rodents and marsupials. And in colugos

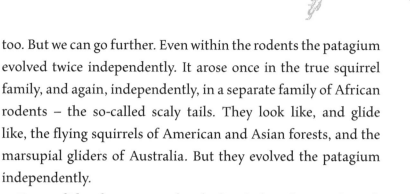

too. But we can go further. Even within the rodents the patagium evolved twice independently. It arose once in the true squirrel family, and again, independently, in a separate family of African rodents – the so-called scaly tails. They look like, and glide like, the flying squirrels of American and Asian forests, and the marsupial gliders of Australia. But they evolved the patagium independently.

Forest gliders have to gain height first before they can launch into their controlled descent. In the forest they do it by climbing a tree. But there are other ways of getting high enough to launch into a glide. There are cliffs. These are favoured by humans with hang-gliders (and a lot more nerve than I possess). Also by many seabirds who can flap their wings but prefer to glide from a cliff if possible, because it's less work and also because there are useful up-draughts around cliffs. Swifts, though consummate specialists in the art of powered, flapping flight, are incapable of taking off from the ground. On the rare occasions when they have to land (in order to nest) they always choose to do so in a high place from which they can launch themselves into the air. Shearwaters in Japan have been filmed by David Attenborough's BBC film crew, queuing up to climb a ramp (a sloping tree trunk) to reach a favourite launch site.

But there's one especially important way in which gliding birds can be carried high, sometimes very high indeed, before launching into a downward glide: *thermals*. Hot air rises. A thermal is a vertical column of rising warm air surrounded by cooler air. Thermals commonly occur because the sun heats the ground unevenly. Some patches, for example rocky outcrops, get hotter than the surrounding land. This heats the air above the warm patch, which therefore rises as a thermal. Cold air comes in to replace it at the bottom of the thermal, is heated in its turn, and rises. At the top of the thermal the air gets cold and descends round the sides of the thermal, circling down again to complete the *convection cycle*. Fluffy cumulus clouds, like puffs of cotton wool, often form at the top of a thermal where it is cool and water droplets condense. These clouds can be spotted from a long way off as tell-tale signs of a thermal.

Now, just as a colugo can mount a tree and launch itself into a glide down to the base of a distant tree in the forest, a vulture or other soaring bird can do the same with thermals instead of trees. But whereas a tree is only tens of metres high, a thermal can lift a vulture thousands of metres. You can see them circling over the African savannah, slowly climbing as they do so, spiralling upwards. The circling helps them stay inside the vertical column of the thermal. Human glider pilots do the same. One of our leading experts on bird flight, the late Professor Colin Pennycuick, was also a pilot and he circled his glider high among soaring vultures, condors and eagles in order to study them.

GAINING HEIGHT FOR THE LONG GLIDE 🐦

Gliding down from one thermal to another. (Obviously not to scale.)

HANG-GLIDER

Is this what it felt like to be a giant pterosaur?

I've never tried piloting a glider and I think I'd like to. Even more inspiring might be hang-gliding where you can steer intuitively by shifting your weight in the harness. I imagine that experienced hang-gliders feel the wing almost as if it were a part of the body. Perhaps that's what it's like to be a gull, wheeling and soaring in the up-draught of a cliff? Or an eagle surveying the savannah from the heights of a thermal. Or even a pterodactyl. But I don't think I'd dare try it. Certainly I wouldn't jump off a vertical cliff as some hang-gliding enthusiasts do. For no very good reason it seems worse to me than jumping out of a plane with a parachute. When I visit the famous Cliffs of Moher in Western Ireland I have to get on my hands and knees to approach the edge, and am tempted to lie flat on my stomach.

We can fancifully think of the savannah as a greatly spaced-out 'forest' of thermals. The 'trees' of rising hot air can be thousands of metres taller than the trees climbed by a flying squirrel, colugo or phalanger. And they are vastly more distantly spaced from their neighbours. So whereas the colugo can glide a horizontal distance of 100 metres or so, the vulture can climb to such a height that its shallow glide from the top carries it many miles, potentially across to the bottom of the next thermal. There it can climb again in preparation to glide down to the base of the next one. Glider pilots speak of thermals as arranged in 'streets'. By steering from thermal to thermal along a street they can stay aloft indefinitely as they journey across country. Eagles and storks use streets in the same way.

How do they know where the next thermal is? Presumably the same way glider pilots do: look for the cumulus clouds that sit atop the thermals; or look for distant columns of circling birds; or read the lie of the land.

Of course, coasting to the next thermal in a street is not the main reason why a vulture wants to gain height. As we saw in Chapter 2, high soaring allows them to scan for food over a very wide area and they glide down when they spot it. Like many birds, they have acute long-distance vision. They can spot a lion kill from miles away, and they also notice when crowds of other vultures are floating down from their thermals to home in on a target on the ground. After feeding on a kill, well-fed and heavy, they need to take off again. For this they have no choice but to flap their wings, costly in energy though it is, in order to get off the ground and reach the base of a thermal.

Dolphins and penguins leap out of the water when swimming fast. This is perhaps an energy-saving ploy, since air resistance is less than water resistance (although other benefits have been suggested). Many fish leap into the air too, as a means of escaping fast-swimming predators such as tuna. When a whole shoal of small fish do this, they land in what looks and sounds like a shower of rain. Some fish, the so-called flying fish, carry their leaps further by using their greatly enlarged fins as wings. They don't flap but glide, sometimes (with the aid of up-draughts from the waves) an astonishing 200 metres, and at speeds of up to 40 miles per hour before touching water again. Although it's true that they don't flap their wings like a bird, some flying fish, while taking off, roll the whole body side to side, which may have a similar effect to flapping the wings. Fish swim by sinuous movements of the tail. As the flying fish takes off, the last thing to leave the water is the still-swimming tail. Sometimes when landing, the fish prolongs the glide by swishing the elongated lower prong of the tail to gain speed and take off again without fully immersing the body.

ON THE ROAD TO MANDALAY,
WHERE THE FLYIN' FISHES PLAY

*I'm actually a little surprised that true flight (staying aloft
indefinitely) hasn't evolved in fish. Give it another few million years?*

As far as a pursuing tuna is concerned, the flying fish has suddenly ceased to exist. The phenomenon called *total internal reflection* means that, from below, the predator can't see its rocketing prey after it breaks through to the air. It has (not quite literally) disappeared into another dimension, like pressing the hyperspace button in a computer game.

Unfortunately for the flying fish, although it may have suddenly vanished from the tuna's world, it equally suddenly shoots up into the world of waiting birds, frigate birds, for instance. Frigate birds can fish from the surface, but they get much of their food by piracy, stealing fish on the wing from other birds. A flying fish must look – to a frigate – much like a bird in possession of something worth stealing. The skill needed to catch one, or to rob a flying gull, must be similar. And frigates are indeed adept at catching flying fish in the air. Frigate birds are black, often with a flash of red, looking like a cross between a prehistoric pterodactyl and the devil. Not for nothing did David Attenborough describe the poor flying fish as caught between the devil and the deep blue sea.

When my sister and I were children, our father composed for us a vivid F-themed monologue about the plight of flying fish: 'Full forty furlongs from Faroes' furthest far-flung frosty foreshore, fifty-five flying fish fled frantically for freedom from forty-five ferocious feathered fowls, flying fishes' fearfullest foe. Forty feet further: flop. Forty feet further: flop.'

Fortuitously forgot felonious frigates, Father?

Squids are also capable of rapid swimming, and some of the faster ones have independently and convergently evolved the flying fish habit, again to escape predators, with the interesting difference that these molluscs swim, and fly, backwards, achieving their great speed by jet propulsion. They force a powerful jet of water out of the mouth, and they speed, like the arrow they resemble, up into the air. They can travel 30 metres or more before landing again some three seconds later in the sea.

It has been convenient to separate gliding from powered flight and give them chapters of their own. But the distinction is somewhat blurred. Even birds that habitually soar in thermals and glide down to the next thermal in a street do flap their wings sometimes. So do albatrosses. In the next two chapters we turn to true powered flight where the power – whether muscle power in a bird or an internal combustion or jet engine in a plane – works continuously to keep aloft indefinitely.

POWERED FLIGHT AND HOW IT WORKS

THE INGENIOUS FLYING SOLDIER

But why a 'soldier'? Surely this wondrous machine
could be put to better use?

CHAPTER 7

POWERED FLIGHT AND HOW IT WORKS

So far we've seen how a large surface area can keep you up with little effort and little expenditure of energy, by gliding, soaring or floating. But if you're prepared to work hard, many more opportunities for defying gravity open up. There are two main ways. The first way is to push yourself directly up. This is the direct and obvious method and it's what helicopters, rockets and drones do. Hovercraft ride on a cushion of air made by downward-facing propellers behind a skirt or curtain. Vertical-take-off jet planes direct a jet downwards to lift the plane off the ground. Stunt flyers like the spectacular 'flying soldier' who flew above Paris on Bastille Day 2019 do something similar.

Leonardo da Vinci was ahead of his time in many ways and his designs include a kind of forerunner of the helicopter. Unfortunately, it couldn't possibly have worked, and not only because it relied on human muscle power. Human muscles are too weak to lift the inevitable weight of man and machine. Modern helicopters have powerful engines burning copious quantities of fossil fuel to drive their great, thudding rotors. The angled blades drive a strong wind downwards, directly pushing the helicopter upwards.

Helicopters also need an additional propeller in the tail, facing sideways (or something equivalent), to prevent the whole craft from spinning like a top. Leonardo seems to have overlooked this last point. The Harrier jump jet, and its successors, don't need it, because they have no rotor. They get their lift by deflecting jet nozzles directly downwards in order to push the plane up off the ground. Having lifted itself off the ground, the plane then directs its jets backwards in order to fly forwards. Then it gets its lift from the wings like any normal plane. And how do normal planes get their lift? That's more complicated and we'll deal with it now.

Unlike helicopters, normal planes obtain their lift by moving forwards fast. They drive themselves forwards with propellers or jets. And the airflow rushing at and past the wings has the effect of lifting the plane in two ways, both of which are important for living flyers as well as manufactured aircraft. The obvious and most important of the two ways is called the Newtonian way. The speed of the plane results in a wind pressing against the wings, and it lifts them because of their slight upward tilt as the plane speeds forwards. You can feel the effect if you stick your hand out of the window of a fast-moving car. Tilt your hand a little upwards and feel your arm pushed up (don't do this if there's any danger that your hand will be mistaken for a signal to another

*Even if the four men ran full pelt round the capstan, this device
wouldn't rise a centimetre above the ground.*

car). So that's the obvious explanation for how wings work: the
Newtonian way. It's the main way planes get their lift. It would
work even if the wings were flat boards angled slightly upwards,
so we could also call it the flat board way.

But there's something less obvious going on as well. There's
a second way in which wings provide lift when driven forwards

fast. The second way is named after Daniel Bernoulli, an eighteenth-century Swiss mathematician. Many people, even some textbook writers, don't fully understand how the two ways work together. Fortunately, planes still stay up even if it's hard to explain, in simple terms, all the complications of exactly how they do.

So, here's the second way, the Bernoulli way, in which wings provide lift. You'll have noticed that the wings of a modern airliner are not flat boards, they are artfully shaped. The leading edge is thicker than the trailing edge. And the shape of the wings in cross-section is a carefully crafted curve, designed to obtain lift as the air rushes over the wing surface, using the Bernoulli principle.

Bernoulli's principle states that when a fluid ('fluid' means gases as well as liquids) is moving across a surface, the pressure on that surface decreases. My attempt of an explanation of it is at the end of the chapter. It's why the shower curtain is sucked in towards you and feels all clammy. It's to combat this that you are often provided with a second curtain outside the bath's rim. The Bernoulli flow in this case is a downward wind generated by the falling water. Now imagine that you had two downward-facing shower heads, one on each side of a curtain. One shower delivers water faster than the other. According to Bernoulli's principle the curtain will be 'sucked' towards the faster shower stream. (I put 'sucked' in quotation marks because what we think of as suction is really higher pressure from the other side.)

Wind, of course, is what a plane wing experiences as it whizzes forwards through the air. Planes get added help by taking off, whenever possible, into the prevailing wind. Now here is the subtle part. According to Bernoulli's principle, the strength of suction depends upon the shape of the surface that the wind is racing past. The air moves faster over the curved upper surface of the wing than over the flatter lower surface. Remember the lesson of the curtain hanging between the fast shower and the slow shower. So, in the same manner as for the shower curtain, the wing is sucked upwards because of the lower pressure on the upper surface.

Exactly why the curved top of the wing causes the air to move faster is quite complicated. It used to be said that two air molecules setting off simultaneously from the front towards the back of the wing, one above and one below it, have to arrive, for some mysterious reason, at the back of the wing at the same time as each other. In other words, those travelling over the curved upper surface have further to go, so – it was thought – they have to go faster. But that's wrong. They don't, as a matter of fact, arrive at the back of the wing at the same time as each other. And there's no reason why they should. Nevertheless, the air molecules do hug the curved upper surface rather than flying off at a tangent, they do travel faster over the curved upper surface than the flatter lower surface, and the Bernoulli effect really does provide a certain amount of lift in consequence.

Having said that, I must again stress that the Bernoulli

STALLING AIRCRAFT
Turbulence patterns in a stalling aircraft.

contribution to lift is normally less important than the first of the two effects I mentioned, the 'flat board' or Newtonian effect. If Bernoulli lift were the most important contributor, planes couldn't fly upside down. And they – small ones, at least – can.

I said the air molecules 'hug' the curved upper surface and don't fly off at a tangent. But that's only true up to a point. If the angle of attack is too high – if the wing is tilted up too steeply – the 'hugging' breaks down, the air molecules break away from flowing smoothly over the wing and they spiral away into horrible turbulence patterns. Bernoulli suction is destroyed, the aircraft suddenly loses lift and is said to stall. Stalling can be dangerous and the pilot must take steps to regain lift, by reducing the angle of attack (usually by tilting the nose down a bit) to restore the proper, smooth flow of air over the top of the wing.

I mentioned 'angle of attack', and we now need to define it and a few other technical terms connected with flying. The angle of attack is the angle of the wing relative to the airflow. Don't confuse it with 'pitch', which refers to the angle relative to the ground. When your plane is taking off, the pitch is high, which is why if you've disobeyed the rules by having a drink on your tray table it tends to spill into your lap. In this case the angle of attack, too, is high. But high pitch doesn't necessarily mean high angle of attack. A fighter plane climbing nearly vertically has a high pitch but a low angle of attack because the airflow rushing over the wing is nearly vertically downwards.

'Pitch' is a verb as well as a noun. An aircraft is said 'to pitch' when its angle relative to the ground tilts down or up. It is said to 'roll' when one wing tilts down while the other tilts up. Pilots control roll with hinged *ailerons* at the back of the wings. They control pitch with similar horizontal hinged surfaces on the tail. Just to complete these three important definitions, a plane 'yaws' when it swivels to left or right. Pilots control yawing by means of a vertical rudder at the back of the tail. Animal flyers, of course, also pitch, roll and yaw.

So far I've mainly discussed planes with fixed wings because the theory is easier with fixed wings. Even so it's still hard. The Wright brothers, and several other early aircraft designers, used 'wing warping', an ingenious system of strings and pulleys with which they could distort the shape of the left wing or the right, thereby steering the plane. Nowadays, wing warping is superseded by hinged ailerons. With bird wings, theoretical calculations of how they obtain lift and forward thrust are harder than with fixed-wing planes. Not only can birds flap their wings. Their wings continuously change shape, in sensitively adjusted ways – a form of wing warping, I suppose. Both flapping and shape-changing make the mathematics of bird flight very difficult to deal with in full detail. We can say, however, that the same two ways of obtaining lift – Newtonian and Bernoullian – work for bird wings as well as plane wings, but in more complicated ways. We'll return to this. Meanwhile, back to the problem of stalling, which applies to birds as well as to planes.

Aircraft use cunning devices to lessen the risk of stalling. Among these is the 'wing slat'. Wing slats are like little extra wings cunningly placed in the front of the main wing to leave adjustable gaps, called slots. Through the slots, the slats deflect to the upper surface of the main wing some extra air that would otherwise have gone elsewhere. This pushes back the critical point at which turbulence starts, pushes it backwards across the upper surface of the wing. It forestalls stalling (sorry). Wing slats allow a steeper angle of attack before the stall is triggered. In normal flight the slats are neatly folded away. Pilots call them into action during take-off and landing, when the angle of attack is at its steepest and the plane is flying its slowest. Modern

PLANES AND BIRDS HAVE TO FACE THE SAME
RULES OF PHYSICS
They come up with similar but not identical solutions.

airliners also sometimes add a graceful twist to the tip of the wing. This reduces turbulence and drag, and is something bird wings do too.

It isn't only planes that can suffer from stalling. Birds are living aircraft, and they are not exempt. Do they have wing slats like planes? Kind of. Many soaring birds have prominent gaps between the feathers near the wingtips, which seem to do the same job. Vultures and eagles show this beautifully. The very large primary feathers at the outer edge of the wing fan out, leaving prominent gaps between them. Each of those big primary feathers acts as a kind of miniature wing or wing slat. This may be especially important for birds spiralling up in a thermal, which is a narrow chimney of warm air surrounded by cold. A vulture needs to circle tightly to avoid straying outside the thermal. The outer wing therefore travels faster than the inner wing, which hence provides less lift and is in danger of stalling. The splayed wingtip feathers are especially useful here, serving as wing slats for the wing nearer to the middle of the thermal.

Engineers frequently perfect plane wings by trying out designs (often miniature replicas) in a wind tunnel. The replica doesn't rush through the air, but the same effect is achieved by the wind in the tunnel blowing past the stationary plane or wing. They sometimes attach little strips of cloth on the top of the wing, to see what is going on, especially what happens to the turbulence when you do things like change the shape of the wing, or alter the angle of attack. When the model wing starts to stall, the strips of cloth rise up just like the feathers on the backs of the stalling egret's wings. Often, testing in a wind tunnel is an easier way to improve a design than mathematical calculations which, in the case of turbulence, are formidably difficult. And it's certainly a safer and cheaper way than building and test-flying a series of whole planes with different wing shapes. Of course, bird wings have been perfected not by somebody doing sophisticated calculations, and not by trial and error in a wind tunnel, but by trial and error in real life. And 'error' means something much

CONTROLLED STALL IN A BIRD

Birds not only can stall, they sometimes deliberately use stalling to help them come down when landing. When a large bird such as a heron or an egret comes in to land, you can see the effects of stalling turbulence lifting the feathers on the backs of the wings.

LEONARDO'S INGENIOUS ORNITHOPTER
It might work as a hang-glider. But flapping the wings with human
muscle power would be ineffective.

more serious in real life than it does in a wind tunnel. It can mean
sudden death. Or, less dramatically, it lowers the expectation of
life and reproduction.

Inspired by birds, Leonardo da Vinci designed a number
of aircraft that look a bit like modern hang-gliders. He also
designed 'ornithopters', aircraft with flapping wings to be
powered by human muscles. Like Leonardo's helicopter, none
of these flapping ornithopters could have worked, although his
gliders might. Flapping flight needs more energy than human
muscles can deploy. Man-powered flight had to wait until the

late twentieth century and the development of ultra-lightweight construction materials, to compensate for the relative weakness of our muscles. When man-powered flight finally came, not surprisingly, the machines didn't flap their wings and they only just barely managed to stay aloft.

The most spectacular is probably the *Gossamer Albatross* designed by Paul MacCready, a brilliant inventor whom I was privileged to meet at his home in California.

> He explained to me, then, his enthusiasm for stream-lining. One of his campaigns was about cars, and the unfortunate way they are designed to *look* streamlined to please would-be purchasers, but usually aren't. In particular, the undersides of cars are not streamlined, perhaps partly because they aren't visible and therefore their appearance doesn't help with salesmanship. Streamlining is immensely important to swimming and flying animals. If you've ever watched swimming penguins or dolphins, either in the wild or in an aquarium, you've probably envied their speed. Human swimmers, even smooth-shaved Olympic champions, seem downright sluggish by comparison. With a mere flick of the tail, a dolphin shoots forwards, almost as though super-lubricated, through the water. And that's not far from the truth. Not only is their body shape superbly streamlined, dolphins continuously shed skin in the form of a kind of dandruff, replacing the outer layer every two hours.

This has the effect of reducing the tiny vortices which would otherwise slow the animal down.

Back to the *Gossamer Albatross* – it was powered by an experienced cyclist pedalling a modified bicycle to drive a propeller and it successfully flew from England across the Channel in 1979. Only just, however. The pedalling pilot was pushed to the limit of a fit young man's endurance, and he almost collapsed within sight of the French coast. The aircraft travelled at between 7 and 18 miles per hour, only a few metres above the waves – as befits the name *Albatross*. Like the Wright brothers, MacCready gave his *Albatross* an extra stabilising wing ahead of the main wing, and his propeller faced backwards. Also befitting the name, the wings were very long and narrow, spanning nearly 30 metres; and it was extremely light, only 98 kilograms, more than half of which was contributed by the cycling pilot's own weight.

MacCready pared from his aircraft every last gram of unnecessary weight. Even the glue used to cement the plane's

GOSSAMER ALBATROSS

*Cycling across the English Channel, Gossamer Albatross only just
managing to keep the cyclist's weight aloft. Flying is energetically very
costly. Right up at the limit of what human muscles can achieve.*

parts together was of a special, ultra-light kind: weight was that critical! Flying animals, too, are as light as they can get away with. Bird, bat and pterosaur bones are hollow: once again there will be a trade-off between making bones as light as possible on the one hand, and hard to break on the other. You might not think teeth weigh much, but it could be that birds lost their ancestral teeth because they were heavier than the horny beak that replaced them. The faster the aircraft the more important streamlining becomes. If you're curious as to why this is, it's because the air resistance goes up as the square of the velocity. It's no accident that modern high-speed airliners, whether designed in America, Europe or Russia, all look the same. It isn't explained only by industrial espionage. Engineers in all countries have to contend

with the same laws of physics. There wasn't the same uniformity of design in earlier years when planes were slower.

After the *Gossamer Albatross*, Paul MacCready went on to other flight projects such as the *Solar Challenger*, a solar-powered aircraft. The *Challenger* was, again, ultra-light and ultra-streamlined. It had solar panels all over its wings and tail, driving a rather large propeller. It could fly at 40 miles per hour and could reach a height of more than 4,000 metres. Later solar-powered planes achieved feats such as flying right round the world. Not in a single hop (for human reasons – the journey took weeks). They did, however, fly at night as well as by day, the batteries being charged by the sun during the day.

The *Gossamer Albatross* pushed the limits of what can be achieved with human muscle power. It achieved what Leonardo's machines aspired to but failed to do, and not by flapping its wings like a bird, as Leonardo's ornithopters were designed to do. Muscle power in the *Gossamer Albatross* drove the craft forwards using a propeller or airscrew. Lift was obtained indirectly from this forward movement.

Powered flight using the internal combustion engine began with the Wright brothers in 1903. Jet engines followed in the 1930s. Astonishingly, only about four decades elapsed between the Wright brothers' pioneering achievement and the first supersonic flight. And only two decades later, members of our species were slung to the moon and back. I use 'slung' advisedly. Rockets set off in an easterly direction, taking advantage of the

Earth's rotational velocity to sling them into orbit. The European Space Agency launch pads in French Guiana are well sited to take advantage of this, being close to the equator where Earth's rotation is best placed to give the rockets a kick-start into orbit.

☞ **By the way,** in case you're wondering how Bernoulli's principle actually works, here's a very simple explanation, without mathematical symbols. First we need to understand what air pressure means at the molecular level. Pressure on a surface is the summed-up drumming of trillions of molecules. Air molecules are constantly whizzing about in random directions, changing direction when they bounce off something such as each other – or such as a surface. When you blow up a party balloon, the inside surface is under more pressure than the outside. There are more air molecules per cubic centimetre inside than out, therefore each square centimetre of rubber suffers more molecular bombardment on its inside surface than on its outside surface. The wind in your face, too, is molecular bombardment. Hold up a card, red one side, green the other. On a still day, the card is bombarded by molecules at the same rate on both sides. But if you hold the card up so the red side is facing into a wind, the rate at which molecules bombard the red side increases, and you can feel the pressure of the wind pushing the card. That's straightforward enough. But now for Bernoulli's principle, which is a little more tricky. Turn the card so it is horizontal, red side uppermost, and the wind is now blowing *across* (both surfaces of) the card. The molecules of air are still bouncing at random off everything including each other and both surfaces of the card. But the movement of the molecules, although still largely random, is now biased in the direction of the wind. That means fewer molecules bombard both surfaces – they are whizzing past the card instead. This is the same as to say the pressure on both surfaces decreases: the card neither rises nor falls. Finally, perhaps with a pair of hairdryers, we contrive matters so that the wind blows faster across the red surface than across the green surface. The pressure on the red surface will decrease more than the pressure on the green surface and the card will rise.

POWERED FLIGHT IN ANIMALS

CHAPTER 8

POWERED FLIGHT IN ANIMALS

Animal flight is more complicated and difficult to understand than that of human machines. This is partly because flapping wings drive the animal forwards (the plane principle) at the same time as they thrust air downwards (more like a helicopter). If you watch a flying bird in a slow-motion film (you need slow motion to even hope to see what is going on) you'll notice that the wings are not simply flapping up and down. The curvature of the wings combined with the supple bendiness of the feathers propels the bird forwards, which in turn provides lift by the two methods we saw in Chapter 7, the Newtonian and Bernoullian methods. At the same time, the downstroke of the wings provides lift in its own right, as we saw at the beginning of Chapter 7, the helicopter part. The upstroke doesn't spoil things by having the reverse effect that we might naively expect. This is partly because of the curvature of the wing, and partly because it twists during the upstroke, and the elbow and wrist joints pull inwards so the area of the wing is reduced compared to the powerful downstroke.

Lacking propellers or jets, birds and other flying animals use their wings to propel themselves forwards as well as to provide lift directly. This is unlike man-made planes, whose wings provide lift but not forward propulsion. The opposite extreme, where wings are exclusively concerned with forward propulsion but not lift, can be seen in penguins, but that's underwater of course, rather than in air. Penguins are buoyant, lighter than water, so they don't need wings for lift. Instead, they use their wings for 'flying' underwater. Unlike on land where they waddle with a slow, ungainly gait, penguins streak through the water extremely fast like dolphins, although dolphins propel themselves in a different way, using up and down movements of the tail. Both dolphins and penguins are beautifully streamlined. Streamlining would have come easily to the ancestors of penguins, already streamlined for flying through the air.

Other seabirds such as puffins, gannets, razorbills and guillemots also use their wings to fly underwater. But unlike penguins, they use their wings to fly through the air too. The best wing shape for air is not the same as the best for water. For underwater flying, smaller wings are better. Puffins and guillemots must settle for a compromise, whereas penguins, who abandoned the air, could perfect their wings for water alone. Puffins have smaller wings than they ideally should have for flying through the air, and consequently have to resort to a very rapid, and energy-expensive, wingbeat frequency. At the same time, their wings are larger than they should ideally be for swimming. Once again we see the evolutionary principle of compromise.

Cormorants propel themselves underwater with their enlarged feet, with only a little help from the wings, which are mostly reserved for flying in air. The great auk, an extinct guillemot/razorbill relative, couldn't fly and, as with penguins, its wings were perfected for swimming. The great auk is sometimes known as the 'penguin of the north'. Indeed, its Latin name is *Pinguinus*, but penguins are not closely related to it. Its wings were small, too small to fly, and very like a penguin's. It's as though the great auk's ancestors were northern razorbills who said, 'Oh, to hell with trying to fly in both air and water. The compromise is too expensive. Let's forget about air and concentrate on water. Then we can do one thing really well.'

It's sad that you and I only just missed the privilege of seeing a great auk. They were driven extinct, as so often, by humans, and as late as the nineteenth century. Maybe, just maybe, our grandchildren will see a great auk. Its genome has already been sequenced from a specimen in a museum in Copenhagen. A colleague of mine is now discussing the possibility of one day using new gene editing techniques to edit the genome of a razorbill, then inserting cells into the gonads of, say, a couple of geese and hatching a great auk from one of their eggs.

Back now to flying through air. Forward propulsion by wings is achieved by a kind of rowing through the air. Hummingbirds

THE PENGUIN OF THE NORTH

Alas, the great auk was driven extinct in the nineteenth century.

HUMMINGBIRD HAWK-MOTH

*When you see this moth and hear its whirring wings
you might think it is a hummingbird. The hummingbird
hawk-moth plies the same trade as a hummingbird,
so has evolved convergently towards it.*

go to the extreme of a rapidly buzzing (humming), sculling movement, in which the wing is turned almost upside down during the upstroke. The wing works almost as efficiently on the upstroke as the downstroke, and it enables hummingbirds to hover like a helicopter and fly backwards, sideways and even occasionally upside down. Hovering was an important evolutionary discovery for birds. Previously, insects had a monopoly in nectar because they could perch on flowers. Birds were too heavy until they invented hovering. Sunbirds are the Old World equivalent of New World hummingbirds. Only some of them can hover. Some flowers are furnished with special projections which seem designed as perches for sunbirds. Among insects hoverflies are champion hoverers. A number of moth species called hummingbird hawk-moths are also good at hovering while sucking nectar from flowers with their immensely long tongues. They get their name from their resemblance to hummingbirds: another nice example of convergent evolution. Dragonflies, too, are good at hovering, which is probably how they earned an alleged pidgin English name, 'Helicopter Belong Jesus'.

Watching a bird in flight, even a slow-motion film, it's hard to disentangle the pushing-down 'helicopter' component from the thrusting-forward 'aeroplane' component. Individual birds shift emphasis from one to the other, for example by emphasising the 'helicopter' component (assisted by a jump) when taking off, then emphasising the 'plane' component when level flight takes over. And different species of bird specialise in one component or the other. Hummingbirds are not the only 'helicopter' specialists. The pied kingfishers of Africa and Asia are the largest birds that can truly hover for sustained periods. Where other kingfishers perch in order to scan for fish, pied kingfishers do it from the air, hovering like a giant hummingbird. Big wings don't hum though.

Kestrels, when looking for prey, hover in a different way, and some purists prefer not to call it hovering at all. What kestrels do is face into the wind and fly at the same speed as the wind but in the opposite direction. This means that their ground speed is zero while their airspeed (speed relative to the oncoming wind) is fast enough to give them lift. Pied kingfishers and hummingbirds – like helicopters – don't need a wind in order to hover.

Birds have separate muscles for the downstroke and the upstroke of the wings. The large breast muscles (*pectoralis major*, 'bigger pec') power the downstroke. These muscles can account for 15 or 20 percent of the body weight. And, as we've already seen in our speculations about Gabriel and Pegasus, they need a large breastbone or keel for their attachment. You might think that the upstroke muscles would have to be above the wing, and so they are in bats. In birds, however, they aren't. They (the *supracoracoideus* muscles) sit below the wing and they pull the wing up by means of a 'rope' (tendon) and 'pulley' over the shoulder. Other muscles twist the wing's angle, and yet others change the wing shape by bending the wrist and elbow joints.

I could have dealt with albatrosses in Chapter 6, the chapter on gliding, because they do mostly soar and glide close to the surface of the sea. But they exploit principles which had not yet been explained, so it's convenient to deal with them now. Albatrosses are masters of energy-economical flight. By the end of its life an albatross may have flown more than a million miles, circling the southern globe again and again. Instead of thermals, albatrosses exploit natural wind currents over the sea to obtain lift. They glide low, in some cases for hundreds of miles without landing, hardly flapping their wings and expending very little energy. The largest species is the wandering albatross of the southern oceans, and it circles the globe continuously, always in one direction, exploiting the prevailing wind. An albatross can't just let itself be blown passively by the wind, for that would give it no lift. It needs the equivalent of a thermal to gain height before the next glide down. So it alternates between gliding downwind, and then turning into the wind. When facing into the relatively slow wind near the sea surface, it becomes like a plane getting lift by the Newton and Bernoulli methods. This pushes it up to a height from which it again turns to glide downwind, blown by faster wind higher up. During this phase of its cycle it loses height like a vulture coming out of a thermal, or a colugo sailing down from a treetop. When it gets down near the sea surface where the wind is slower, the albatross turns into the wind and climbs again. It repeats this cycle indefinitely. It also skilfully adjusts its flight surfaces to take advantage of eddies and up-draughts caused by the waves. These wave-generated up-draughts are less consistent than thermals, more irregular. Exploiting them requires sensitive moment-to-moment adjustment of the flight surfaces, which can be achieved only by sophisticated 'electronics' – an advanced nervous system.

For an expert but very large glider like an albatross, taking off is a problem. Albatrosses can flap their wings but, as usual, flapping flight is energy-expensive and, for large birds, very laborious. When taking off from land they do pretty much what a plane does. They run fast along a 'runway', into the wind, until they get enough airspeed to pull their wings upwards. Albatross breeding colonies really do have recognisable runways like planes. I've seen these in the Galapagos and New Zealand. Unlike a plane they also flap their wings to gain added lift. When out at sea, although they can glide over the waves for prodigious distances, they do sometimes land, for instance, in order to fish or perhaps rest. Again, taking off is a problem. They flap their wings as hard as they can and run fast over the surface, resembling the strenuous take-off of an old-fashioned Sunderland Flying Boat (but with the important addition of flapping wings). Swans are also large enough to have the same problem of laboriously taking off from water. I regularly hear the loud, rhythmic swish of their wings and I rush to watch them rise, slowly and with great effort, from the surface of the Oxford Canal just outside my window.

SWANS ON THE OXFORD CANAL

Big birds struggle to take off. But they do.

☞ That birds can run over the water surface may seem surprising, **by the way**, but it is not uncommon. As we saw, bird wings are stiffened by feathers rather than only bones. This means they aren't tethered to the hind legs like the wings of bats and pterosaurs. Bird legs are therefore free to run. Many birds have powerful legs and can run very fast, ostriches up to 45 miles per hour. It is their strong legs that enable some birds to run on the surface of water. Lizards are distantly related to birds and some basilisk lizards, such as the aptly named Jesus Christ lizard of South and Central America, skitter across the water surface on their strong hind legs at 15 miles per hour, nearly as fast as they can run on land. Western grebes of North America have a magnificent and rather comic courtship dance in which the male and female birds run in tandem across the water, so fast that only their feet and tail touch the surface. It is a similar, though more laborious, ability that albatrosses employ in their take-off run on the surface of the sea. Albatrosses have large webbed feet, which must help. Grebe feet are not exactly webbed, but each toe has leaf-like lobes which achieve much the same effect.

Insects were undisputed masters of the air for nearly 200 million years before vertebrates, in the form of pterosaurs, joined them. I wonder why it took the vertebrates so long. I'm accustomed to thinking that, if there is an open niche (way of

life or trade) some animal will swiftly evolve to fill it. It's hard to see why the many open flying niches – escaping from predators, searching for food from the air, migrating long distances, catching insects on the wing, all the things we talked about in Chapter 2 – were not filled by vertebrates much earlier. In the light of Chapter 4, I suppose it could have been their small size that enabled insects to take to the air so early.

In the Carboniferous period some 300 million years ago when most of our coalfields were laid down, there were outsize dragonflies with a 70-centimetre wingspan flitting – if 'flitting' is the right word for such a behemoth – among the giant ferns and clubmosses.

You might notice an amusing little error in Michael Crichton's science fiction thriller *Jurassic Park*. The adventurers encounter dragonflies with a metre-wide wingspan. The author seems to have been carried away by his story, and to have forgotten the clever basic idea of it, which was that Jurassic Park's scientists bred creatures from DNA in blood sucked by mosquitoes that later became embalmed in amber. But mosquitoes don't suck blood from dragonflies, and anyway the oldest amber-preserved insects lived 100 million years later than the giant dragonflies of the Carboniferous era.

It's been suggested, with evidence from several sources, that the gigantism among Carboniferous dragonflies was possible only because there was more oxygen in the atmosphere in those days. Perhaps as high as 35 percent according to upper estimates, compared to today's 21 percent. The insect system of piping air through the whole body, rather than to specialised lungs, works efficiently only with a relatively small body. A more oxygen-rich atmosphere would have pushed that size limit up. With higher oxygen levels, forest and prairie fires (ignited by lightning) would have been more common. Perhaps the giant dragonflies used their big wings to escape ubiquitous fires. They'd have had better luck than their crawling contemporaries, the giant Carboniferous millipedes, 2.5 metres long, or *Pulmonoscorpius*, a giant scorpion 70 centimetres long – for me, such stuff as nightmares are made on. As for *Eryops*, to call it a giant newt might make it sound relatively harmless, but it was a voracious carnivore, up to 3 metres long, the Carboniferous occupant of the crocodile way of life.

Insects have no bones. You can get a better idea of their skeletons by looking at their larger relatives such as lobsters. Instead of bones they have a set of horny, jointed tubes, the *exoskeleton*, housing the soft, wet parts of the body. Insect wings are not modified arms like bird wings, they're papery outgrowths of the exoskeleton, hinged at the wall of the thorax. The muscles that raise the wings pull down on the near end of the wing inside the body wall, so the wing goes up like a lever. In a few large insects such as dragonflies, the downstroke is achieved by muscles on the farther side of the hinge, as you'd expect. But in a greater number of insects the downstroke is achieved in a less obvious way. Muscles running along the thorax contract,

causing the roof of the thorax to bow upwards. This has the indirect effect of levering the wings downwards – hinged as they are on the thorax.

Insects can achieve astonishingly high wingbeat frequencies, 1,046 times per second in the case of some midges: that's two octaves above middle C. It's a version of that infuriating noise you hear when a mosquito is about to bite you – what the poet D. H. Lawrence called its 'hateful little trump'. As you can imagine, it would be hard to achieve such frequencies by nerves alternately telling the wing muscles 'up-down-up-down-up-down' 1000 times per second. And they don't. Instead, these insects have *oscillatory muscles* that spontaneously vibrate. It's a sort of high-speed shiver. The flight muscles of a gnat or mosquito or wasp are little reciprocating engines that are either on or off. Instead of alternating instructions, 'up-down-up-down', the central nervous system says simply 'fly' (switch on the oscillating engine). And then, after a while, 'stop flying' (switch off the engine). There's no throttle, no accelerator. All the time when it is switched on, the muscle engine vibrates at a fixed frequency which is determined by the 'resonant frequency' of the wings. It's as though the wing is a pendulum, swinging at a fixed frequency but much much faster than the pendulum of any clock. As you'd expect from the pendulum comparison, the wingbeat frequency rises dramatically if you shorten the wings by amputation. Admittedly the note we hear seems to change as a mosquito whines around your ear, or a bumblebee

GIANT WATER BUG
The largest insect with an
oscillating wing-muscle engine.
Don't mess with those jaws!

buzzes around a flowerbed. But that's mostly because, as the insect changes direction, what are called inertial effects change the 'pendulum's' behaviour. On a much slower scale, this is why Harrison's marine chronometer was such an important breakthrough. Pendulum clocks are inaccurate on a rolling ship.

Some larger insects like dragonflies and locusts are quite different. As in birds, each upstroke and each downstroke is separately commanded by the central nervous system. It's mostly small insects that use the oscillating engine type of muscle. But not entirely. Probably the largest to do so are the giant water bugs ('bug', by the way, although commonly used to refer to any insect and even to bacteria or viruses, is actually a strict zoological term limited to insects that suck their food, members of the order Hemiptera). Giant water bugs are formidable tropical creatures with alarming jaws that can deal a painful, though not venomous, bite. Although they mainly live in water they can fly, and they use their oscillating flight muscle engines to do so. Because of their large size they were adopted by my Oxford professor, 'Laughing John' Pringle (so nicknamed because he seldom cracked so much as a smile), for his studies of oscillatory muscle. It's hard to see what you're doing if you try to do research on the muscle fibres of a gnat.

Bats, the only true flying mammals, flap their wings in a similar way to birds. But, while their wings lack the helpful curvature provided by feathers, bats seem to have a different trick up their leathery sleeves. In addition to the main muscles controlling the flapping of the wings and the spacing of the fingers between the webbing, there are rows of thin, thread-like muscles embedded in the skin of the wings. I don't know whether these *plagiopatagiales* (I don't know how to pronounce it either) are derived in evolution from the muscles that all mammals have in their skin for erecting the hairs (the ones that give us goosebumps when we are cold – a fascinating relic of the time when we had enough body hair to keep us warm). Whatever their origin, it looks as though they are used to adjust the tension in the different parts of the bat's flight surfaces. Also, perhaps, to produce curvature in a different way from bird wings. These fine-adjustment muscles within the skin combine with the finger movements' larger-scale adjustments to provide sensitive control of the flight surfaces. Such sophisticated control is likely to be important in a fast-flying hunter like a bat. Indeed, bristling as they are with high-tech radar (actually sonar) instrumentation, bats remind me of high-performance attack fighter planes. Small bats, that is. Large fruit bats, including flying foxes, don't need high-speed manoeuvrability since they don't pursue moving targets like the insect-hunting small bats. Fruit doesn't run away.

> ☞ Unlike the little bats, the large fruit bats have large eyes. And no sonar; or it's poorly developed and done in a different way – suggesting convergent evolution. In appearance, fruit bats remind me of pterosaurs, although they are, of course, mammals. Did pterosaurs have sonar? Some had large eyes, suggesting that they flew by night but probably relied on vision. Incidentally, I've also wondered whether ichthyosaurs, extinct dolphin-like reptiles, had sonar. Dolphins have highly sophisticated sonar, evolved entirely independently from bats. But ichthyosaurs, unlike dolphins, had very big eyes so they probably didn't have sonar.

Aircraft have to cope with a trade-off between stability and manoeuvrability. The great evolutionist and geneticist John

Maynard Smith was an aircraft designer during the Second World War before he went back to university to become a biologist ('deciding that aeroplanes were noisy and old-fashioned'). He pointed out that the trade-off is important for living flyers like birds, just as it is for man-made planes. Very stable aircraft can pretty much fly themselves, or at least a relatively inexperienced pilot can handle them. But the trade-off is with manoeuvrability. Stable planes are no good as military fighters, which need to be agile and nimble in the air, swift to turn and dodge. Highly manoeuvrable planes are unstable – there's that trade-off again. They can only be handled by expert pilots with swift reflexes. And even expert pilots nowadays would be helpless without on-board computers in very advanced planes. The day may come when pilots, however expert, are completely superseded by electronic guidance systems.

On-board computers, and expert pilots, need instruments – equivalent to sense organs and aids to sense organs. In the animal kingdom, flies, especially hoverflies, are spectacularly manoeuvrable, and they have superb instrumentation. Unlike other insects, all flies (which range from gnats and mosquitoes to large crane flies or 'daddy longlegs') have only one pair of wings (hence the Latin name, Diptera). The second pair of wings has shrunk, over evolutionary time, to become *halteres*, little sticks with a knob on the end, behind the remaining wings. Halteres are flight instruments. They whirr as though they were miniature wings, but they are entirely the wrong shape and much too small for flying. Instead, they serve as a kind of gyroscope to help with steering and stability. If its halteres are removed, the insect can't fly; it's too unstable. You can make it a stable flyer again by gluing on a tail made of a little feather of the kind trout fishermen use for tying flies.

John Maynard Smith pointed out that early pterosaurs such as the Jurassic *Rhamphorhynchus* had an extremely long tail with a sort of oar on the end. It would have been a stable flyer but poor at manoeuvring. Compare it with *Pteranodon* of the late Cretaceous, 100 million years later. It had almost no tail at all. According to Maynard Smith it would have been manoeuvrable but unstable. It probably relied on 'electronics' – sensitive control of flight surfaces by the brain – to compensate for the lack of a stabilising tail. Did *Pteranodon*, perhaps, have muscles in its wing membranes like modern bats? If anything, it might have had greater need of them because pterosaurs, having only a single finger in the wing, lacked the fine finger adjustments available to (equally tailless) bats. And did *Pteranodon* have a more sophisticated brain than *Rhamphorhynchus* to cope with the necessary 'electronic' control? How did it use that great rear projection from the skull, balancing the forward-protruding jaws? Did the whole head, perhaps, function as a front-end rudder, automatically steering the creature in whatever direction it chose to look?

No modern bird has a long bony tail like *Rhamphorhynchus*. What we typically call the tail of a bird is made of feathers

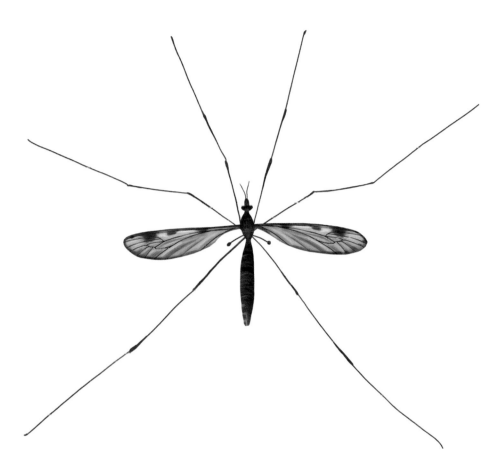

CRANE FLY ('DADDY LONGLEGS')
AND ITS 'GYROSCOPES'

Most flying insects have four wings but flies have only two (hence the name Diptera). The second pair of wings evolved to become sense organs called halteres, little sticks with a knob on the end, which work like tiny gyroscopes.

TWO PTEROSAURS 100 MILLION YEARS APART
Rhamphorhynchus *(top)* had a long tail which would have made
it a stable but unmanoeuvrable flyer. Pteranodon *(bottom),*
a late pterosaur, had hardly any tail and was probably unstable
and manoeuvrable.

without bone, while the true tail is the stubby 'parson's nose' of the roast chicken. But *Archaeopteryx*, the famous Jurassic fossil which probably was close to being an ancestor of all birds, had a long bony tail like most reptiles, including *Rhamphorhynchus*. It presumably was aerodynamically stable, and unmanoeuvrable in Maynard Smith's sense.

One of the reasons birds need to be manoeuvrable is that they often fly in dense flocks, and it's important to avoid colliding with neighbours. As to why they flock, there are various reasons. Perhaps the most important is safety in numbers. Predatory birds normally take only one prey at a time, and predators are usually well spaced out, occupying separate hunting territories. The larger your flock, the smaller the chance that you'll be the one caught by the local hawk or eagle. The 'safety in numbers' effect works especially well if you can contrive to place yourself in the middle of the flock rather than the edge. This advantage applies to schooling fish too, and herding mammals. Such groups can be very large, even hundreds of thousands strong, and the risk of collision is surely high.

Winter flocks of starlings – they have a name of their own, *murmurations* – are exceedingly numerous with hundreds of thousands in any one murmuration, and they display spectacular feats of coordination. They wheel and climb, dive and turn, seemingly in unison, almost as though the whole gigantic flock is a single organism. This illusion is enhanced by the fact that the edges of the flock are sharply defined: it seems to lack stragglers that are neither properly in the flock nor out of it. After their amazing aerial dance, quite suddenly, like a noisy rainstorm, the birds plummet down to their night-time roost.

The watcher is tempted to suspect a leader – a master choreographer – but there is none. Each individual bird follows the same simple set of rules, keeping an eye on its nearest neighbours, and coordination emerges as a result. This has been mimicked by computer simulation, and it's a fascinating example of how computer modelling can illuminate our understanding of reality. Starting with the pioneering Boids model by Craig Reynolds, computer programmers have adopted the following important principle. First program a model of a single bird, building into it simple rules for how to respond to its neighbours, keeping them at particular angles, for example. Then make hundreds of copies of this one bird. Finally, look at what happens when all these hundreds of copies are released in the computer. What these model birds do is 'flock', in a most realistic manner, exactly like real birds. It's important to understand that Reynolds and his successors didn't 'program a flock'. They programmed one bird. Flocking then *emerged* as a result of cloning up many copies of that one simulated bird. This principle of 'emergence' is hugely important in biology generally. Complex organs and behaviour emerge when each of many small components follows simple rules. Complexity is not built in: it emerges. But that's a great topic which deserves a book in itself.

'AS OF INNUMERABLE WINGS'

A murmuration of starlings is one of the wonders of the world.

CRANES IN V-FORMATION
Apart from the one at the front, all get the benefit
of the slipstream from the one ahead.

Back to why flocking is a good thing for birds. Although the main point is probably to baffle predators, here's another rather subtle benefit, which applies not to murmurations but to the familiar V-formation of many travelling birds. They position themselves to make use of the slipstream from a bird in front. The best position for this is diagonally behind – hence the V-formation of flying geese, storks and many other birds. Of course, the bird at the front of the V doesn't get the advantage. It's been shown in ibises that they take turns for the difficult hot spot in front. Racing cyclists employ the same trick; and so do military aircraft to save fuel. The Airbus company is investigating the possibility of formation flying for large airliners to save fuel.

Yet another benefit of flocking is taking advantage of others discovering food. However good your eyesight is, a flock has more eyes and one of them may spot a good source of food that you have missed. There's experimental evidence that great tits watch each other feeding, and even do copycat searching in the kind of place where fellow flockers are seen to have found food.

What's the next solution to the problem of obtaining lift? Be lighter than air.

CHAPTER 9

BE LIGHTER
THAN AIR

MONTGOLFIER BALLOON

A work of art in the sky.

CHAPTER 9

BE LIGHTER THAN AIR

Planes, helicopters and gliders, bees and butterflies, swallows and eagles, bats and pterosaurs are so-called heavier-than-air flying machines. Balloons and airships are lighter-than-air machines. As their name suggests, they float without effort, held up by a gas such as hydrogen or helium which is lighter than air, or by hot air which is lighter than the surrounding cold air. More accurately they are held up by the heavier air falling around them, lifting them by Archimedes' principle. As far as I know, lighter-than-air flying machines exist only as products of human invention. I know of no true animal balloons.

In the history of human technology, lighter-than-air flying devices long pre-date heavier-than-air ones. The first human flight was in Paris in 1783: a hot-air balloon made by the Montgolfier brothers. Joseph-Michel Montgolfier had noticed something curious about laundry drying over a fire. Pockets of hot air pushed the garments towards the ceiling. This inspired Joseph-Michel to team up with his business-minded brother

Jacques-Étienne to make hot-air balloons. They built a series of increasingly large balloons, experimenting with animal passengers before they risked human life. Aristocratic life, as it happened, for the first human flight was by the Marquis d'Arlandes, together with Pilâtre de Rozier. De Rozier was a scientist, and a resourceful one for, according to one report, the balloon caught fire and he staunched the flames with his coat.

Only days later, the first human flight in a hydrogen balloon was launched, also from Paris. This flight was by Professor Jacques Charles, he after whom Charles' law, governing the expansion of gases, is named. Charles rode in a beautiful boat-shaped car slung beneath his balloon. He landed a few miles outside Paris and was met by two dukes at a gallop. Not content with his maiden flight, Charles promptly took off again, having promised the Duke of Chartres that he would return. Which he did. Fortunately this hydrogen balloon didn't catch fire, for that would have been goodbye balloon and goodbye intrepid aeronauts. These early ballooning exploits were fraught with danger, and several early aeronauts did indeed lose their lives. De Rozier himself ended tragically, if predictably, when he later set off in a hybrid balloon of his own design, a fire balloon suspended under a hydrogen balloon. See what I meant by 'predictably'?

The Montgolfier balloon of de Rozier's earlier ascent was a thing of beauty, fit for the royal personages who were among the thousands of spellbound witnesses. Modern hot-air balloons are made in a colourful range of funny shapes, as well as the more conventional pear-shape. The first Montgolfier balloons were tethered. It's hard to work out the details from conflicting contemporary accounts, but it seems that they left their fire behind on the ground when they flew, and presumably landed again fairly swiftly when the air in the balloon cooled. Later Montgolfier balloons carried a brazier underneath them and the aeronauts fed the fire with straw. Modern hot-air balloons burn gas from a propane cylinder, which fires short precision blasts of intense heat deep within the interior of the balloon.

You might think the ideal lighter-than-air machine would contain a vacuum, for what could be lighter than that? Unfortunately, in order to resist crushing by the air pressure outside, the craft would need a massively strong shell made of something like steel, whose weight would defeat the purpose, to put it mildly. A workable balloon or airship has to have a light envelope and be filled with gas lighter than the nitrogen/oxygen mix which is our planet's air. Hydrogen is the lightest element of all, and it, or coal gas which is rich in hydrogen as well as other light gases such as methane, was used in early airships. Bad idea! Hydrogen is highly – explosively – flammable. Mindful of the tragic destruction of the giant Hindenburg airship in 1937, airship designers now favour the second lightest gas, helium, instead.

☞ **By the way**, in the quantities needed to carry people, helium is expensive, but you can buy small cylinders of the stuff for filling party balloons. It's non-flammable and relatively harmless. It's also good for further amusement at the party. This is because an additional effect of its being lighter than air is that sound travels through it nearly three times faster than in air. This means that if you breathe helium into your lungs you'll sound like Minnie Mouse. Don't overdo it. Too much helium, or helium inhaled too deeply, can be bad for you.

Nowadays, given the cost of helium, hot-air balloons are much more common. Hot air, as we've seen in connection with thermals, is lighter than cold. Heating the air in a balloon with a roaring blowtorch is cheaper than filling it with helium, although it is rather noisy, which spoils some of the charm of floating over the quiet countryside. Of the three balloon trips that I have enjoyed, one was with a television crew. I was supposed to wax eloquent about the gentle charm of evensong in English country churches as we drifted past their towers and steeples. Not surprisingly the crew had to limit their filming to the lulls between roaring blasts from the propane burner.

The world of professional balloonists seems to be a small one. My third and most memorable trip was in Burma and, by sheer coincidence, my pilot turned out to be the very one who had piloted me and the film crew over the parish churches of

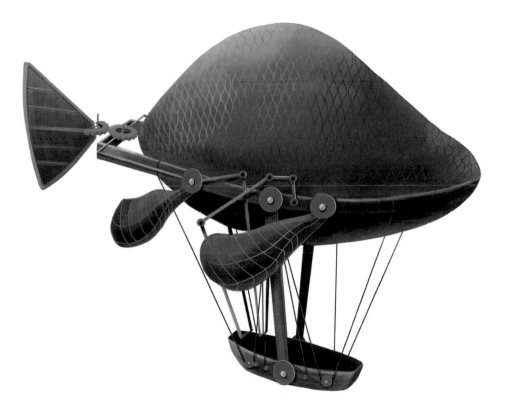

DIRIGIBLE FISH BALLOON DESIGN
Evolutionary intermediate between balloon and airship?

the peaceful English countryside. In Burma we drifted over a truly spectacular landscape dotted with literally thousands of Buddhist temples and pagodas shrouded in the early morning mist over the Plain of Bagan. A sight to see before you die.

Balloons, unlike airships, are hard to steer. Airships amount to large balloons with a cabin slung underneath, and with propellers to drive them horizontally. You can steer them – hence the name dirigible, which literally means steerable. Early balloon designs employed steering devices modelled on boats, including sails, rudders, oars and paddles. I suppose they were the first dirigibles but I doubt, looking at them, that they were very dirigible.

In a simple balloon, all you can control is your height. You can attempt to locate a level where the wind happens to be blowing in your preferred direction, which is a pretty hit-or-miss way to steer. To increase height in a hydrogen or helium balloon you throw out some of the ballast (such as sand) which you have thoughtfully taken with you in the basket. In a hot-air balloon you switch on the propane burner for a quick blast. To sink, you pull a rope to open a vent at the top of the balloon which releases some of the hot air – or gas, as the case may be. It's surprising how sensitive the balloon is to slight changes in weight. When using ballast, you need throw only a small amount overboard in order to gain altitude. This is because the balloon is an *aerostat*, in equilibrium with the air that surrounds it. What does this mean?

The density of the atmosphere decreases with altitude, so there'll be some critical height where a balloon hangs in perfect equilibrium. If the balloon is lower than its equilibrium altitude it will rise. If it finds itself higher than its equilibrium altitude it'll sink. Throwing out sand (or firing up the blowtorch) has the desired effect by changing the balloon's 'preferred' – that is, equilibrium – altitude. As another example of this, balloonists sometimes use a simple but clever device for automatically regulating their altitude, which works only when the balloon is close to the ground. Dangle a long rope, the 'trail rope', out of the basket. The weight of the rope, slight as it is, is significant. When the balloon is low, most of the rope is on the ground, so its weight is not part of the craft's net weight. If the balloon rises, more of the trail rope is above the ground and its weight tends to pull the balloon down a little. In this way the trail rope automatically regulates the balloon's height. I find this surprising – you'd think a mere rope would be too light to make a difference – but it just goes to show what a sensitive aerostat a lighter-than-air machine is. Shortly before the giant airship Hindenburg exploded on that terrible day in New Jersey in 1937, it sank lower than it should have. Film shows the crew frantically trying to gain height by jettisoning water ballast, and it doesn't seem to be very much water compared to the size of the aircraft itself. On the first balloon crossing of the English Channel by Jean-Pierre Blanchard in 1785, he and his American companion were obliged to jettison everything in their beautiful boat-shaped car, including even their clothes, for the same reason.

I earlier mentioned my old boss the serious 'Laughing John' Pringle, and his research on the oscillatory flight motor of insects. He, incidentally, had been a champion glider pilot, so he knew a thing or two about staying aloft. So did Sir Alister Hardy, his beaming predecessor as Oxford's Linacre Professor of Zoology, who had been a balloon enthusiast in the 1920s. Hardy wrote a delightful little book, *Weekend with Willows*, describing an eventful, indeed hazardous balloon journey by four young gentlemen from London to Oxford, piloted – in somewhat reckless fashion – by the celebrated aeronaut and airship designer Captain Ernest Willows, who later died in a tragic ballooning accident. Their balloon was lifted with coal gas, and Hardy describes their initial quest to find a London gasworks willing to pump the necessary gas for them. The flight from London to Oxford was immortalised in a 426-line epic poem by one of the party, Hardy's friend Neil Mackintosh. I'll quote just seven couplets to convey his wit and the spirit of adventure of the expedition – something of the same spirit, it seems to me, as *Three Men in a Boat*, a comic Victorian story of a boat trip up the Thames to Oxford by a group of similar young men and a dog called Montmorency.

STRIPPING FOR DEAR LIFE

Blanchard's 1785 Channel crossing was a triumph. But, dangerously losing height, he and his companion had to jettison everything from their gondola, including even their clothes and their steering oar.

At one point somewhere between London and Oxford – Hardy and his friends had no idea where – there loomed out of the mist...

> *An unforeseen and deadly trap*
> *Which might have caused a grave mishap.*
> *'Grave' is a word appropriate,*
> *For just before it was too late*
> *We saw emerging from the gloom,*
> *Encompassed by grave and vault and tomb,*
> *A church upon a hill so high,*
> *Its steeple seemed to touch the sky.*
> *And as with fear we did perspire,*
> *Lest we be spiked upon the spire,*
> *The ballast bags we quickly manned;*
> *The graves received a little sand*
> *Instead of corpses, crushed and gory,*
> *Else you'd not have heard my story.*

The trouble with balloons, as we've just seen, is that you can't steer them. You never know where you are going to land, so – as I know from personal experience when ballooning in the countryside around Oxford – you have to have a retrieval crew chasing you in a vehicle. My own return to earth on that Oxfordshire trip was made eventful by an unexpected last-minute gust of wind which blew us sideways, dragging us bumping through a hedge and across two fields until we finally tumbled out of the basket. I inadvertently made a soft landing on a charmingly uncomplaining young woman of the party. Also with us in the basket was a visiting Japanese professor with a limited command of English. The farmer who owned the field came rushing over as we were picking ourselves up and dusting ourselves down. 'Where have you come from?' he excitedly asked. This was a question the professor had met before and he knew the answer. 'Ho,' he replied without hesitation, 'from Japan!' In Alister Hardy's more laid-back times, they didn't have a back-up crew with a car and trailer, as we had. Balloonists would watch for a convenient railway line below, and land beside it. Having packed the balloon away into its canvas bag they'd flag down the next train, which would obligingly stop and pick them up, no doubt to the bemused delight of the delayed passengers.

As I said at the beginning of this chapter, it seems that no non-human animal has evolved anything truly equivalent to a lighter-than-air balloon. Small spiders and caterpillars do something called 'ballooning'. It's also called 'kiting', which is a better name because it doesn't involve being lighter than air. The spider releases threads of silk, which act as a kind of kite, catching the wind and lifting the little spider into the air. Some spiderlings travel hundreds of miles in the so-called aerial plankton, which we'll meet in Chapter 11. There is evidence that ballooning spiders obtain some lift, when taking off, from the Earth's electrostatic field. You can observe static electricity yourself. Rub a piece of plastic on your hair. You'll then notice that the plastic attracts small objects, little scraps of paper, say. It isn't magnetism, although it looks a bit like it. It's static electricity. And it is a static electric force that some baby spiders use to launch themselves into the air.

But how about true ballooning? Do any animals actually float by being lighter than air? A naturally evolved balloon doesn't seem completely out of the question. The separate ingredients are not unknown around the animal kingdom. Some man-made balloons have been constructed of silk, which is both light and strong. Silk, of course, was invented by spiders and also independently by insects, notably the caterpillars which we call silkworms. Some caddis larvae make silk traps, fishing for small crustacean prey and, unlike typical spider webs, these have a weave that looks close enough to make a balloon. Silk weaving, then, is an available technology. But what gas could they use to fill the envelope? It's hard to imagine how animals could evolve the ability to manufacture helium. Some bacteria can make hydrogen, and there's talk of exploiting them commercially to make fuel. Animals exploit bacterial expertise in other spheres, for example to generate light. Methane, another light gas, is readily generated in animals. Methane emitted by cows, again actually made by bacteria (and other micro-organisms) in their stomachs, is a worrying source of greenhouse gas in the atmosphere. It's also produced by decaying vegetation. Known

WOVEN SILK

This trap made of silk by a caddis larva is not a balloon.
But it shows that animals are capable of making one of the
components that would be necessary in a balloon.

as 'marsh gas' it sometimes catches fire as a 'will-o-the-wisp'.
As for hot air, the most impressive example I know of animal
heat production is the weapon used by some Japanese bees

against marauding hornets that raid their nests. They will mob a hornet, surrounding it in a tight ball of bees. By vibrating their abdomens, the bees raise the temperature to 47 degrees Celsius. This literally cooks the hornet to death. If it cooks some bees to death as well, that doesn't matter: there are plenty more to take their places. However, although some separate components of balloon technology – heat, hydrogen, methane and tightly woven silk fabric – seem to be available through natural evolution, I don't know of any examples where they have been brought together to obtain lighter-than-air lift-off. Who knows, maybe it still awaits discovery?

⌁ Water is much denser than air, so the watery equivalent of lighter-than-air flying is easy and commonplace. We do it every time we swim. Konrad Lorenz begins his account of snorkelling by recalling his childhood dreams of flying. We are mostly made of water anyway, and the air in our lungs makes us lighter still. Sharks are slightly heavier than water and, like flapping birds in air, they need to keep swimming in order not to slowly sink. But teleost fish (bony fish as opposed to cartilaginous fish like sharks) deserve an honorary mention in this chapter because they are finely controlled *hydrostats*, capable of expertly varying their density. In this respect they are like dirigible airships, which are finely controlled aerostats. As we saw before, an

aerostat finds its level at an altitude where the lift provided by the less dense gas exactly balances the weight of the craft, including passengers. It then hovers at equilibrium. A fish does the same kind of thing, by fine control of its *swim bladder*. The swim bladder is a bag of gas deep inside the fish. By changing the amount of gas in the bladder, the fish can change its own density and so rise or fall to find a new level where it is again at equilibrium. This is why teleost fish seem to drift so effortlessly. It's one reason fish tanks are such restful spectacles to have in a room. The swim bladder allows fish to expend only the energy they need to propel themselves horizontally. Unlike flying birds and unlike sharks, teleost fish don't have to spend energy on lift. Birds could do the same thing in the air if they had an aerial swim bladder filled with methane. But they don't.

Fish aren't the only animals to have evolved something equivalent to a swim bladder – a means of regulating their density. Cuttlefish – despite their name, not fish but molluscs, relatives of squids and octopuses – maintain hydrostatic equilibrium by withdrawing or injecting fluid into their porous 'bone' – the cuttlefish bone that is typically given to caged birds to supplement their calcium.

As a means of flying usefully, lighter-than-air craft have huge limitations, which is why dirigible airships are nowadays such a rare sight in our skies. They are for fun, or for advertising stunts, rather than for commercial transport. Even hydrogen, the lightest gas, is not sufficiently lighter than air to lift a heavy load unless an enormous volume of the gas is used. The necessarily large envelope of the gas container must itself be light, which means it will be flimsy and vulnerable, often consisting largely of soft fabric with minimal support from a rigid or semi-rigid skeleton. The stable shape for a bag of gas under pressure is a sphere. That's why most balloons, from Montgolfier on, are spherical or approximately spherical. But a sphere is not a good shape for travelling fast through the air, so advanced dirigibles propelled by engines, such as the famous Zeppelins, tend towards the streamlined cigar shape. But the further the dirigible departs from the stable spherical shape, the more its gasbag needs rigid skeletal support to retain its shape. This adds extra weight, which increases the need for an even larger volume of gas just to lift the airship itself, let alone cargo or passengers. And the more voluminous the gasbag, the greater is the drag as it tries to move forwards through the air. If speed is what you want, dirigibles can't begin to compete with heavier-than-air planes deriving their lift from their horizontal movement.

On the other hand, because they don't consume fuel in order to obtain lift, dirigibles are cheap to run. So if you don't care about speed, perhaps because you are transporting cargo with no crucial delivery time, you might be tempted to use a dirigible. But since the maximum speed of a dirigible is so slow – the world record only just tops 70 miles per hour – it can't cope with the sort of headwinds that a big jet plane takes in its stride. Presumably they could go faster but they'd need great big engines like a jumbo jet. And those engines would be too heavy to lift using the aerostat principle.

CHAPTER 10

WEIGHTLESSNESS

FALLING AROUND THE WORLD

The astronaut feels as though he's flying but he's really in free fall.

CHAPTER 10

WEIGHTLESSNESS

Let's turn now to our final method of defying gravity, weightlessness. On the face of it, it's one that is used only by humans. And technologically advanced humans at that. If you're an astronaut in the Inter-national Space Station (ISS), you enjoy a wonderful illusion of flying. These fortunate but rare individuals come closer than anyone to realising the dream of Leonardo. In the space station you have no sensation of 'up' or 'down'. No surface of your living space deserves the title of floor or ceiling. You float like a ghost, and when you have dinner (perhaps from a toothpaste tube, because food would float off a plate) with a companion, each may seem to the other to be upside down. To pass from one room of the space station to another you fly, pulling yourself through the air by handholds. If you jump up from what you temporarily regard as the floor, no matter how gently, you'll fly 'up' to the 'ceiling' and bump your head. If astronauts need to go outside to do maintenance or repairs they, again, float freely, and need to be tethered for fear of

becoming separated from the spacecraft. They drift effortlessly like a balloon, or like a fish in perfect command of its swim bladder. Unlike a fish, however, the reason they float is not that they have the same density as the surrounding medium. Far from it. Their surrounding medium inside the space station is air, outside it is a near vacuum, and they are far denser than either. Why, then, do they float?

We come now to an error so common that we must deal with it promptly. Many people imagine that astronauts are weightless because they are far from Earth and out of reach of its gravity. Wrong, wrong, wrong! The space station is not all that far from Earth – closer than London is to Dublin – and Earth's gravity is tugging it almost as strongly as if it were at sea level. No, the astronauts are weightless in the sense that, if they sat on a weighing machine, it would register their weight as zero. Both astronauts and weighing machines are floating freely around the cabin, with the result that the body exerts no pressure on the weighing machine. So their weight is zero.

Astronaut and weighing machine, space station and everything in it float because they are in free fall. They are continuously falling. Falling around the world. Gravity is still acting on them, pulling them towards Earth's centre. But at the same time they are whizzing at high speed around the planet, whizzing so fast that they keep missing the Earth even while they fall. That is what being in orbit means. The space station in orbit floats for an entirely different reason than the balloon

in aerodynamic equilibrium. The balloon is supported
by the pressure of the air that surrounds it. That
is why balloons don't fall. Astronauts in orbit
do fall. They fall continuously. The moon is
falling and has been falling for more than 4
billion years. Falling around the world, falling
in perpetual orbit.

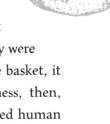

Are balloonists weightless? No, of course not.
Their feet are planted firmly on the floor of the basket
and they have no tendency to drift out of it as if they were
in orbit. If they stood on a weighing machine in the basket, it
would register their full weight. True weightlessness, then,
is our final method of defying gravity. Only advanced human
technology has achieved it. But wait! Is that strictly true? Think
about it this way.

The first astronaut in orbit was Yuri Gagarin, in 1961.
Struggling to catch up, the United States launched Alan
Shepard, also in 1961. He didn't go into orbit but went into what
amounted to an extremely high jump, more than 100 miles high,
at the end of which he splashed down in the Atlantic. During the
acceleration phase of the flight, Shepard was far from weightless.
If he had sat on a weighing machine he would have registered
6.3 times heavier than his normal weight. He actually
was 6.3 times heavier. After the rocket motors were
switched off, however, which means during the
majority of his upward motion as well as during

much of his descent until his parachutes opened, he and his capsule went into free fall. A weighing machine too, if he had had one with him, would have registered his weight as zero during a major part of his spectacular jump.

Let's now return to our question of whether any non-human animal has achieved weightlessness. Our preliminary answer was no, because none has evolved a rocket motor capable of achieving orbital velocity. We've just seen that Alan Shepard, unlike Yuri Gagarin, didn't achieve orbital velocity. Nevertheless, both men achieved weightlessness. And now, think about that proverbial jumper, the flea, and ask how different it is from Alan Shepard. Lacking a rocket motor, a flea has to use muscles.

> An interesting side issue, **by the way**, is that muscles can't move fast enough to provide the sudden explosive acceleration you need for a leap as high as a flea's. The energy of the flea's (necessarily slow) muscles is stored in an elastic spring. It's the same principle as the catapult or the longbow or crossbow. A catapult can propel a stone at a much higher speed than could be achieved simply by the arm muscles that were used to pull the rubber back. Stretching the rubber stores the muscular energy. Fleas, like other jumping insects such as grasshoppers, are equipped with a wonderful elastic material called *resilin*. Resilin is the equivalent of the rubber in a catapult, but it's better than rubber: it's super-elastic. The flea's muscles 'wind up' the resilin, taking their time

to do so. Then the stored energy in the elastic is suddenly released in both legs simultaneously and the flea springs high into the air.

According to mathematical theory, the absolute height an animal can jump to is not related to its size. In practice, of course, there's enormous variation, as some animals like fleas and kangaroos (and Olympic high-jumpers) specialise in jumping while others, like hippos and elephants (and me), don't. A flea can jump about 20 centimetres high, which is not so different from the standing jump of a person. As a proportion of the flea's body size, however, its jump would be roughly equivalent to a person leaping over the Eiffel Tower. Another example of champion leapers is jumping spiders, charming little fellows who jump by pumping fluid explosively into their hollow legs. Although bigger than a flea, a jumping spider leaps about as high, following the rule that absolute height of jump is independent of size.

Theoretically, if we neglect complicating factors like air resistance, the flea's trajectory, or the jumping spider's, should be a graceful curve, which mathematicians call a parabola. Alan Shepard's trajectory looks pretty much like a larger version of the flea's parabola, except that his active propulsion continued for the first part of his upward journey. The flea's active propulsion stops the moment it leaves the ground. Shepard's trajectory was complicated, too, by various manoeuvres that he

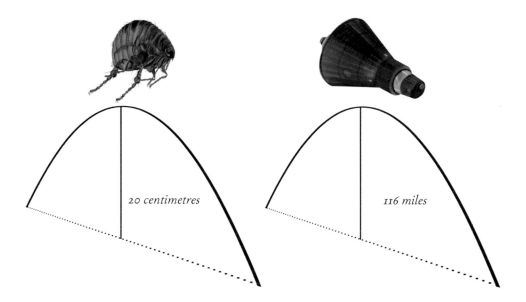

20 centimetres

116 miles

ALAN SHEPARD'S GIANT LEAP

And the smaller but still impressive leap of the flea.
Both are parabolas but with complications.

manually controlled, through the firing of retro-rockets, and by the parachute at the end.

'Let the cow be a sphere and assume it is in a vacuum' is a friendly joke at the expense of theoretical physicists' habit – a perfectly sensible one – of oversimplifying reality to make the calculations easier. Let's follow the joke and gleefully neglect the complicating factors for both flea and Shepard. Both leap in a graceful parabola. The difference is that the flea's high jump is 20 centimetres, the astronaut's 101 nautical miles; the flea is launched by muscular energy stored in resilin, the astronaut by a rocket. Both achieve weightlessness, the flea for less than a second, the astronaut for several minutes. Now imagine that the flea is sitting on a tiny weighing machine. It's hard to imagine a flea-sized weighing machine but we are licensed as physicists to do so. And – again neglecting air resistance and other complications – the flea, and the weighing machine on which it is sitting in their joint free fall, would weigh exactly the same as the astronaut in his: zero.

Now let's bring in Gagarin, or the modern space station, to our theoretical fairy tale. The weightlessness of Gagarin in orbit is no different from the weightlessness of Shepard, or the flea. This includes not just the downward part when they were more obviously falling. As soon as the flea leaves the ground, it is falling, albeit travelling upwards. As soon as Alan Shepard's rocket motors ceased to push him, he was falling (again upwards). And weightless. Gagarin's weightlessness merely lasted longer. The weightlessness of an astronaut in the space station lasts longer still. And the weightlessness of the moon lasts for billions of years. We conclude that astronauts are not the only animals to defy gravity by being literally weightless. 'Even educated fleas do it.'

CHAPTER 11

AERIAL PLANKTON

FLOATING FREE AS AIR

Why are there no giant balloon animals sweeping up the
aerial plankton as whales do in the sea?

CHAPTER 11

AERIAL PLANKTON

High in the atmosphere we encounter the so-called aerial plankton or *aeroplankton*. It's a mixed population consisting of large numbers of pollen grains, spores, flyaway seeds, tiny insects like *Tinkerbella*, miniature spiders trailing little parachutes of silk, and lots more besides. I've already mentioned the 'ballooning' spiders, and there are lots of other tiny animals plus plant and fungal spores, bacteria and viruses up there, too. The name 'plankton', of course, is borrowed from the sea. Like a vast waving prairie, the surface layers of the sea abound in microscopic plants, single-celled green algae and bacteria which harvest the sunlight for photosynthesis, and thereby form the starting point of the food chain. Microscopic animals in the plankton eat the algae and they in turn are eaten by larger creatures and so on. Planktonic creatures in the sea practise what has been called vertical migration: they descend to the depths at night, where they are safer, then migrate up by day to catch the sunlight on which all life depends.

I've already mentioned my old Oxford professor Sir Alister Hardy in connection with a memorable balloon flight from London to Oxford. The major research of his life was on the plankton of the sea.

☞ He invented the Continuous Plankton Recorder. This instrument is towed behind a ship; not necessarily a specialised research vessel, any ship will do. It contains an extremely long band of silk, which is continuously paid out from one reel to another. Sea water passes through the silk, and planktonic organisms are trapped as it does so. When the reel of silk is examined, the position in the sea where each planktonic organism was caught can be calculated – knowing the speed and course of the ship, plus, of course, the speed with which the silk was spooled from one reel to the other.

While researching this book, it came as no surprise to me to find that Professor Hardy also turned his attention to aerial plankton, together with a colleague. Their 1938 paper is a model of clear writing, friendly, almost chatty, in a style that no scientific journal, alas, would accept today. They used two kites to string up a net which caught the aerial plankton. Delightfully, they used an old car, a 1920s Bullnose Morris, as part of the equipment. Having driven to the launch site in the car, they jacked up the rear axle and used one rear wheel with its

SIR ALISTER HARDY

The great expert on marine plankton turned his attention to the
plankton of the air using a pair of kites winched by a jacked-up car.

tyre removed as a winch to reel the kites in and out. Others have
used nets towed from planes for a similar purpose.

Unlike plankton in the sea, aerial plankton is not the main photosynthesis layer sustaining any food chain, although there are algae and green bacteria up there which are capable of photosynthesis. Insofar as plants enter the aerial plankton, they are using the air as their medium of dispersal, including dispersal of pollen and seeds. You might ask why it is so important to spread seeds far and wide. No doubt it's partly to avoid competition between parent and offspring. But there's a more subtle reason. It involves an intriguing mathematical theory and it applies to animals as well as plants. I won't go into the detailed mathematics but will follow my usual habit of trying to explain a mathematical theory in words, without algebraic symbols.

If a plant or animal is living in the best possible place, there would seem to be an obvious advantage in planting its offspring in the same place. What, after all, could give them a better start in life than the best possible place? Nevertheless, the mathematical theory shows that an animal (or plant) that takes steps to send at least some of its offspring a long way away will spread more of its genes, in the long run, than a rival that drops all its offspring right next door to the parent. This is true even if 'right next door' is (at present) the best place in the world and 'a long way away' is on average worse. You can get an inkling as to why this is if you reflect that catastrophes like floods or forest fires occasionally happen and destroy 'the best place in the world'. Of course such catastrophes are rare, and are no more likely to strike the 'best place in the world' than anywhere else. Nevertheless, if you look back through the history of any particular place, however perfect it may be now, you'll probably hit a time somewhere back in the past when it was the victim of a catastrophe.

When thinking about evolution I often find it helpful to look backwards in time, back through long generations of ancestors. One day I plan to write a book along these lines called *The Genetic Book of the Dead*. Every living creature, animal or plant, is the latest in an unbroken line of successful ancestors. The ancestors were successful by definition: they survived long enough to become ancestors – and becoming an ancestor is the Darwinian definition of success. I am using this way of thinking to explain why plants need to disperse their seeds far and wide rather than just drop them on the ground under the parent. And why animals need to send at least some of their offspring out like Christopher Columbus or Leif Ericson, seeking their fortune in unknown lands.

A successful animal (or plant) may live in the same place as its parents, but probably not in the same place as its ten times great-grandparents. It will look back on at least some ancestors who owed their success to having left the parental haven, sent away to seek their fortune in the wild unknown. In the case of plants, 'sent away to seek their fortune' might mean seeds broadcast to the shifting winds.

The majority of those wildly broadcast seeds fell on stony ground and perished. They didn't become ancestors. But any living creature looking back in time will almost certainly number among its ancestors at least some who started life far from their parents and thereby escaped the forest fire, earthquake, volcano, flood or equivalent which unpredictably ravaged the parental home area. This is partly why plants invest so much in spreading their seeds to distant places rather than taking the easy course of dropping them nearby. And the same goes for animals. This, in part, is what the aerial plankton is all about.

My late friend and colleague William Hamilton is famous for his brilliant contributions to Darwinian theory. Some say he was the greatest Darwinian of the second half of the twentieth century. Many of his far-sighted ideas are now universally accepted by biologists everywhere. One of his more minor contributions was the theory I have just tried to explain which, in its mathematical form, he proposed along with another of my Oxford colleagues, the Australian physicist turned biologist Robert May, who was later to become President of the Royal Society and Chief Scientific Adviser to the British Government. But Bill Hamilton also made daring suggestions which are still waiting to be taken seriously and which sound as wild as the winds. Among these is his remarkable suggestion about aeroplankton.

His idea is that micro-organisms such as bacteria and single-celled algae, high in the atmosphere, seed the formation of rain clouds. They have evolved to do this because it benefits them to be carried far across the world and then rained down to begin a new life in a new place. It's one of those ideas that's hard to test, and it's fair to say that not many scientists take it seriously. I wouldn't write it off, not least because it could be seen as an especially grand example of what I long ago called (in a book of that name) 'the extended phenotype'. Bill had a track record of being far ahead of his time, and he was right too often for any idea of his to be easily discounted. The idea was the inspiration for a moving speech on the occasion of his burial.

First, the backstory. Some years before his death, Bill published two versions of a typically strange paper called 'My Intended Burial and Why'. In it, he wrote,

> I will leave a sum in my last will for my body to be carried to Brazil and to these forests. It will be laid out in a manner secure against the possums and the vultures just as we make our chickens secure; and this great *Coprophanaeus* beetle will bury me. They will enter, will bury, will live on my flesh; and in the shape of their children and mine, I will escape death. No worm for me nor sordid fly, I will buzz in the dusk like a huge bumble bee. I will be many, buzz even as a swarm of motorbikes, be borne, body by flying body out into the Brazilian wilderness beneath the stars, lofted under those beautiful and un-fused elytra which we will all hold over our backs. So finally I too will shine like a violet ground beetle under a stone.

As we mourners stood on a grey, cloudy afternoon on the

edge of Wytham Wood near Oxford, scene of so many years of great ecological field research, Bill's much loved Italian partner Luisa Bozzi knelt, grieving, and spoke into the open grave. After saying why it hadn't been possible to carry out his wish to be laid out in the Brazilian forest, she uttered these remarkable words:

> Bill, now your body is lying in the Wytham Woods, but from here you will reach again your beloved forests. You will live not only in a beetle, but in billions of spores of fungi and algae brought by the wind higher up into the troposphere, all of you will form the clouds and wandering across the oceans, will fall down and fly up again and again, till eventually a drop of rain will join you to the water of the flooded forest of the Amazon.

Luisa, herself, sadly died not long afterwards. But her poetic words are carved on a stone bench, standing close to Bill's grave. I have just visited the place again, as I often do. He would certainly have appreciated such a beautiful farewell from the love of his life. So perhaps the clouds had a silver lining, whether or not seeded by aeroplankton.

THE VISIONARY BILL HAMILTON ↜

Greatest Darwinian of my lifetime

Bill, now your body is lying in the
Wytham Woods, but from here you
will reach again your beloved forests.
You will live not only in a beetle,
but in billions of spores of fungi and algae
brought by the wind higher up
into the troposphere, all of you
will form the clouds and
wandering across the oceans,
will fall down and fly up again
and again, till eventually
a drop of rain will join you
to the water of the
flooded forest
of the Amazon.

CHAPTER 12

'WINGS' FOR PLANTS

'SHE LOVES ME NOT'

Each dandelion seed is small enough to fly easily, and it increases its
surface area by means of its own little parachute.

CHAPTER 12

'WINGS' FOR PLANTS

With a few exceptions such as the Venus flytrap and the sensitive plant, *Mimosa pudica*, plants lack the equivalent of muscles. They can't move. Plants do, however, have a powerful need (see Chapter 11) to spread their seeds, and to exchange pollen with other members of their species. The principal medium through which they do both is the air. If plants don't exactly fly through the air, they achieve the equivalent of flying in various indirect ways. So they deserve a chapter in this book.

Thistledown, dandelion puffs and many other seeds are scattered literally to the four winds. They use some of the principles of flying that we have already met. Each dandelion or thistle seed is small, and it has a neat little feathery parachute whose large surface area enables it to float great distances. Sycamore seeds are bulkier: once again we have a trade-off. Very tiny, light seeds like those of dandelions lack the nutrients that get a larger seed off to a good start in life. Sycamores reach a different compromise. Their seed is not small, and fewer seeds

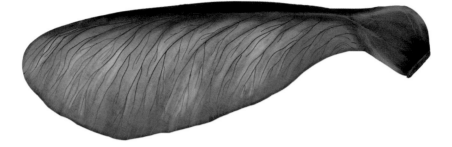

WINGED SYCAMORE SEED

If you didn't know better, might you think it was an insect wing?

are therefore produced: packing a seed with food is costly. And the larger sycamore seed has the necessary large wing to carry it, but not very far. It really does look almost exactly like an insect wing, doesn't it? Of course it doesn't flap. Instead it's blown by the wind, spinning as it descends like a little toy helicopter.

Sycamore seeds are among several that behave like miniature helicopters. But perhaps the most spectacular example of a flying seed is that of *Alsomitra macrocarpa*, the Javan cucumber. Its fruit is a gourd from out of which it launches a succession of beautifully fashioned gliders. Each glider has two paper-thin wings stretching out from a central seed. They soar and swoop as elegantly as any tropical butterfly. Other plants' seed pods are spring-loaded and explode, flinging seeds out at high speed. Seeds of the Common Storksbill then drill into the ground by alternately curling and uncurling their 'awn', a strap-like structure.

JAVAN CUCUMBER SEED

Butterflying down through the forest.

Many plants borrow bird wings (and mammal legs) to carry their seeds far away. Burrs have little hooks like Velcro, which fasten onto fur or feathers and get carried away to be deposited somewhere else. Delicious fruits are designed to be eaten, and the point is not to give the eaters pleasure. The seeds are designed to pass through the gut and come out the other end, well manured. But not all would-be eaters of fruits are equally desirable to the plant. Birds, having wings, are likely to carry seeds further before depositing them, which might well be a good thing for the plant. This may be why deadly nightshade berries, for example, are lethal to most mammals but edible by birds.

Pollen, too, needs to be dispersed. Why? It's important to avoid inbreeding. Exactly what sex is good for is much debated by scientists. Why do most animals and plants mix their genes with those of the opposite sex? Why don't they do as female aphids and stick insects do: make copies of themselves without bothering with males or mating? You might think the answer is obvious but I promise you it isn't. Whatever the reason, it must be a powerful one because almost all animals and plants do sex, immensely costly and time-consuming though it is. And if you mate with yourself it defeats the whole object, whatever that object might be. This is why plants, including hermaphrodite plants with both female and male parts, go to great lengths to get their pollen transferred to another plant. Through the air. So, like seeds, pollen needs to fly.

The simplest way for pollen to fly is just to be blown in the wind. Pollen grains are very small, so, following Chapter 4, they float in the breeze. But this method is rather wasteful. A windblown pollen grain has to be extremely lucky to find the appropriate female part, the *stigma* of another plant of the same species. The low probability of doing so is balanced by puffing millions of pollen grains out, great wafting clouds of them. This is what many plants do and it works well enough.

But is there a less wasteful way of doing it: a different solution to the problem? One idea might immediately occur to your imagination. The plant could develop a little flying vehicle for the pollen, a miniature chariot with wings. The chariot would need the equivalent of sense organs to detect other plants of the same species. And the equivalent of a little brain and nervous system to control the wings and direct the flying pollen-carrier to the right target. Well, it's not a bad idea and it could work. But why bother? The air is already full of little flying vehicles. Bees and butterflies, for instance. Bats. Hummingbirds. They already have fully working wings propelled by muscles, controlled by brains and with sense organs capable of homing in on targets. All the plant need do is find a way to exploit them. Lure insects to pick up your pollen and persuade them to fly the precious cargo where it needs to go.

Maybe 'exploit' is the wrong word. Why not construct a partnership so that both sides benefit? How about paying insects for their services? Pay them with aviation fuel: nectar. Of course plants don't sit down with bees and negotiate the deal: 'I'll give you nectar if you fly my pollen for me. Sign here.' Instead, Darwinian natural selection favours plants that happen to have a genetic tendency to make nectar. Genes for making nectar get passed on via the plant's pollen grains, carried by bees lured by the nectar. Nectar is expensive to make, I should add. Flowers pay dearly for their hired wings.

Insects accidentally pick up pollen which sticks to their bodies while they are sucking the nectar up. The pollen then brushes off on the stigmas of other plants when they are visited for more nectar. It isn't just bees and butterflies, of course. Hummingbirds love nectar too, and so do their Old World and Asian equivalent, the sunbirds. Beetles and bats are pollinators for some plants. Anything with wings is liable to find them borrowed by plants.

How do the bees and butterflies, hummingbirds and others find the nectar? Natural selection favours plants that *advertise*: 'Come and get your nectar here.' Flowers do it partly by alluring perfumes: scents that we find attractive too in many cases, such as roses and lilies. In other cases not so much. Flowers designed to attract some kinds of flies smell like rotting meat.

Bats have wings and some bats like nectar, so it's not surprising to find plants that specialise in hiring bat wings to carry their pollen by night. But since bats use sound echoes rather than light beams to find things, the equivalent of a conspicuous advertising hoarding has to appeal to the ears rather than the eyes. *Marcgravia evenia*, a climbing plant in the Cuban rainforest, has leaves shaped like dish reflectors. The thing about the dish reflectors is that they serve as powerful beacons for echoes coming from many directions. To a bat, living in a world of echoes, the dish-shaped leaf presumably 'glows' like a bright neon sign.

Fascinatingly, there's evidence that flowers and bees generate electric fields which interact with each other and help guide bees into the target when they are at close quarters. There's even some evidence that electrostatic forces suck pollen from male flower organs onto the bee's body, and then repel pollen from the bee's body onto female flower organs.

(01)

(02)

(03)

NARCISSUS, OF LEGEND,

FELL IN LOVE WITH HIS OWN IMAGE

What might he have thought of how a Narcissus *looks to*
an insect – its target audience, after all.
Daffodil as we see it (01), in ultraviolet light where spots appear,
invisible to us (02), and covered with electrostatic dust (03).
Actually an insect probably sees any daffodil as flicker, like a
stroboscope, rather than with the five petals that we see.

But it's mostly through the eyes that flowers attract their pollinators. Insects have good colour vision. So do birds. Both can see a colour outside the range that we can see, ultraviolet, and flowers take advantage of this. Many flowers have patterns of stripes or spots that can be seen only in ultraviolet. Insects can't see red but birds can: which is why if you see a bright red wild flower you'd probably be right to suspect that it is aiming to attract birds. A meadow full of wild flowers is a Piccadilly Circus or a Times Square for bees and butterflies, with brightly coloured petals as the meadow's neon signs. Both the colours and the scents have been enhanced by human gardeners, acting as selecting agents as though they were giant bees.

Hiring the wings of bees, butterflies and hummingbirds targets the pollen more accurately than broadcasting it out into the wind. A bee emerges from a flower dusted with pollen, and heads straight for another flower. But the second flower might not belong to the right species. Is there a better way? Could a flower do something to make sure its pollen will be carried to the same species of flower? Is there some way to reduce 'promiscuity' among insects and encourage 'flower fidelity'? Yes. Flowers have a number of tricks up their Technicolor sleeves. Within any one species, most flowers are the same colour. Insects who have just visited a flower tend to go on to another flower of the same colour. This somewhat reduces the likelihood that pollen will be delivered to the wrong species of flower. Only somewhat, however. What else can be done?

There are flowers that keep their nectar at the bottom of a long tube, so that only insects with a very long tongue can reach it. Or only hummingbirds with a very long beak. The sword-billed hummingbird of South America has a bill longer than its body, a bill so awkwardly long that it can't preen large parts of the body, which must be rather inconvenient. Perhaps more than inconvenient: as we saw in Chapter 5, birds spend a great deal of time preening their feathers, which suggests that preening makes an important contribution to survival. A bird that can't preen its wing feathers might find its flying impaired. In the face of this, the pressure to grow such a long beak in a hummingbird must be exceptionally strong. This remarkable swordbill seems to have co-evolved with the exceptionally long nectar tubes of a particular flower, *Passiflora mixta*. The pink petals advertise the opening of the tube, which extends so far back that only a swordbill can reach the nectar. The flowers can be confident (you know what I mean) that only a swordbill will visit it, and confident that the swordbill will go on to another flower of the same species. Bird and flower are faithful partners of one another. Pollen will not be wasted by being transported to a flower of the wrong species.

There's a beautiful parallel case of a moth. In 1862, while Charles Darwin was working on his orchid book, a Mr Bateman sent him some specimens including the Madagascar orchid *Angraecum sesquipedale*. '*Sesquipedale*' comes from a Latin word meaning as long as a foot and a half. This orchid has an

DRASTIC STEPS TO ENSURE POLLINATOR FIDELITY

Passiflora mixta *keeps its nectar at the bottom of a long tube. It can be
'confident' that only a sword-billed hummingbird can reach it and will
carry its pollen to another flower of the same species. It hires the wings of the
swordbill and only the swordbill.*

'GOOD HEAVENS, WHAT INSECT CAN SUCK IT?'
The answer (though Darwin died too soon to see it) turned out to be
Xanthopan morganii praedicta.

extraordinary nectar tube which can indeed be as long as that. In a letter to his friend the botanist Joseph Hooker, Darwin said, 'Good Heavens, what insect can suck it?' He then made the bold prediction that somewhere in Madagascar there must exist a moth with a tongue long enough to reach deep into this orchid's nectar tube. Darwin died in 1882. Twenty-five years later, an entomologist in Madagascar discovered a local subspecies of the African moth *Xanthopan morganii*. The tongue of this moth can reach 30 centimetres (about 1 foot), triumphantly vindicating Darwin's prediction and justifying its subspecies name, *praedicta*.

Some flowers, especially orchids, go to extraordinary lengths to seduce insects into pollinating them. And I do mean seduce. Bee orchids look like bees, different species of orchid resembling different species of bee. Male bees are fooled into trying to mate with the flower. In the course of their fumbled attempts, pollen sticks to the bee and is flown to the next orchid that the bee tries to mate with. Orchids don't fool only the eyes. Some of them also mimic *pheromones*, strong-smelling chemicals with which female insects lure males to mate with them. Other orchids mimic flies; yet others, wasps of various kinds. Insect-mimicking orchids make no nectar. Unlike other flowers which pay their pollinators, these insect-seducing orchids cheat them, and get their services free.

If scattering pollen to the wind is wasteful because most of it never reaches its required destination, the orchids in this chapter represent the opposite extreme, the 'magic bullet' of pollination with minimal wastage. Somewhere out at the extreme magic bullet end of the spectrum are the hammer orchids, ten species of the genus *Drakaea*, which live in Western Australia. Each of the ten species is pollinated by its own species of wasp, so pollen suffers the least wastage from being deposited on the wrong species of female flower or otherwise lost. Each flower has a dummy female wasp on the end of an 'arm' that has a hinged 'elbow'. They also secrete a chemical that mimics the seductive perfume of a female wasp of the preferred species. Females of these wasp species are wingless. Their normal habit is to crawl to the top of a plant stem and wait to attract a winged male by smell. The male then seizes the female and carries her off in his arms, mating with her on the wing. A male tries to do the same to the orchid's dummy female. He grabs 'her' and attempts to fly off with 'her'. His frantic wingbeats propel him upwards but the dummy female doesn't play: 'she' doesn't let go of the plant. Instead the 'elbow' of the orchid's 'arm' bends up, carrying the male and banging him hard and repeatedly against the *pollinia* (orchids keep their pollen in discrete bundles called pollinia). After a number of bangs the pollinia loosen and stick to the male wasp's back. Eventually he gives up trying to detach the 'female', and flies off to try his luck with another one (when will they ever learn?). The drama repeats itself. Again the male is repeatedly banged back and forth, and this time the pollinia separate from his back and become attached to the stigma of this second orchid. Pollination has been achieved and the wasp has gained nothing for his pains (and perhaps pain).

HAMMER ORCHID, ITS ANVIL LOADED WITH POLLEN

An almost incredibly elaborate device for ensuring that pollen is properly delivered. The male wasp 'thinks' he's found a nice female, tries to fly off with her in his arms, and is banged against the pollen. Repeatedly.

Also at the magic bullet end of the spectrum is the bucket orchid, *Coryanthes*, found in South and Central America. It may be the most complicated of all flowers. By mutual evolution, it has built up a very intimate relationship with a particular group of small, shiny green bees, the so-called orchid bees. Male orchid bees use a pheromone – a very particular sexual perfume, an aphrodisiac scent – to attract females. But they can't make the pheromone unaided. Instead, the orchids make ingredients for them in the form of a waxy material which the bees store in spongy pockets in their legs, so they can use it later for attracting females. While visiting an orchid to collect the waxy materials to make the aphrodisiac, the bee is likely to fall into the flower's 'bucket', which contains a liquid. He swims about trying to get out of the bucket and discovers that the only way out is through a narrow tunnel. As he struggles through the tunnel, two pollinia stick to his back. Eventually he bursts free and flies off, bearing the pollinia. Older but not wiser, he enters another flower, again falls into the bucket, and again wriggles out through the tunnel. This time the struggle scrapes the pollinia off the insect's back and they fertilise the second flower.

> ☞ It's an interesting question, **by the way**, how this arrangement evolved – the plant manufacturing a major component of the pheromone for the bee. I would guess that originally the bees' ancestors made their own pheromone, and the plant gradually took over the role, step by step.

But my favourite candidate for the ultimate magic bullet is the intimate relationship beween figs and fig wasps. I devoted a whole chapter of another book, *Climbing Mount Improbable*, to it. Here I'll simply state that there are more than 900 species of fig and almost every one of them has its own private species of fig wasp which exclusively pollinates it. There's magic bullet for you!

Plants, then, use wings to spread their DNA, just as the owners of the wings use them to spread their own DNA. But plant wings are borrowed wings, borrowed (or hired) from insects, birds or bats. If you're wondering whether there were ever pterosaur-pollinated flowers, so am I. I don't know the answer, but I like the question and like the image that it conjures up. It's not unlikely, because flowering plants evolved in the Cretaceous period when there were still plenty of pterosaurs about.

Strictly speaking, fungi are not plants. They're their own thing, actually closer cousins to animals than to plants. But they don't move around like animals, so it's convenient to think of them as plants. And because they don't move around, like plants they sometimes find it convenient to borrow insect wings. In their case it's to carry neither seeds nor pollen but spores. There are mushrooms that glow a ghostly green in the dark. The light attracts insects which presumably benefit the fungus by dispersing its spores.

CHAPTER 13

DIFFERENCES BETWEEN BETWEEN EVOLVED AND DESIGNED FLYING MACHINES

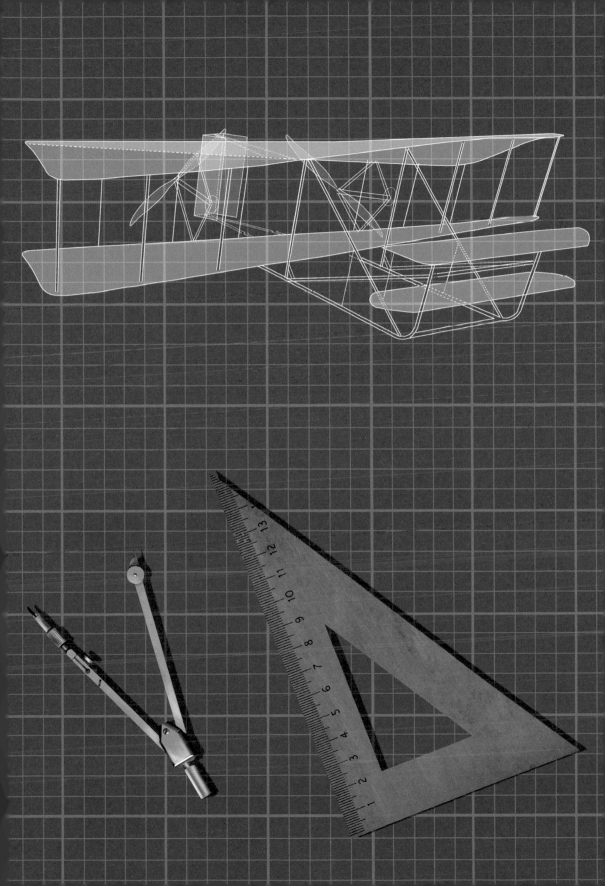

BACK TO THE DRAWING BOARD

Incidentally, the great evolutionist John Maynard Smith was an aircraft designer as a young man before he decided to go back to university to retrain as a biologist.

CHAPTER 13

DIFFERENCES BETWEEN EVOLVED AND DESIGNED FLYING MACHINES

This book has considered about half a dozen (depending on how you count) methods of getting off the ground and staying aloft – defying gravity. In each chapter, where possible I have compared human-designed flying machines with corresponding animal flyers. But the process of *becoming* good at getting off the ground is radically different in the two cases. Animals have become flying machines through millions of years of slow, gradual improvement, getting better at it as the generations went by. Humans have built better and better flying machines through successive designs on drawing boards, and the improvements have taken place on a timescale of years and decades, instead of millions of years. The end results are often similar (and that's not surprising because the problems that need to be solved are the same – the same physics). They are so similar that I might have given the false impression that they come about in the same way. It is time to put that mistake right.

When we face a problem, like how to avoid stalling in a flying machine, it's convenient to think, 'How would I set about solving the problem?' In the case of man-made planes, design engineers really do think like that. They notice a problem. They imagine possible solutions to it, such as wing slats. They sketch ideas on drawing boards, maybe they get together in brainstorming sessions with a whiteboard, or a computer with graphics software. Perhaps they build prototypes, or scale models which they test in a wind tunnel. And the solution that emerges is finally put into production. The whole research and development process (R & D) takes only a few years or even less.

With animals the process is different, and much slower. The R & D, if you can call it that, takes many generations spread over millions of years. No thought goes into it, no clever ideas, no deliberate ingenuity, no creative inventiveness. There are no drawing boards, no brainstorming engineers, no wind tunnels to test prototypes or scaled-down models. All that happens is that some individuals in the population just happen, by random genetic luck (mutation and sexual shuffling of genes), to be a little bit better than average at, say, flying. Maybe a mutant gene gives a falcon a slight edge in speed. Individual falcons bearing this gene are a little bit more likely to catch prey. Or perhaps a mutant starling is a bit more manoeuvrable than rivals in the flock, which makes all the difference between dodging a predator and being eaten. When a starling is eaten because of a 'slow flying gene' the gene is eaten too, and is not passed on to the next generation. Or another genetic type might be a little bit less likely than others to stall, because of a subtle difference in wing shape. So they are a bit more likely to survive and therefore reproduce, passing on the genes that made them slightly better flyers than their contemporaries. Slowly slowly, gradually gradually, generation by generation, the good-at-flying genes become more numerous in the population. The bad-at-flying genes become less numerous, as animals that possess them are a little bit more likely to die or fail to reproduce.

The same thing is going on all the time with lots of different genes in the population, each influencing flying in its own way. So, after lots of generations, after millions of years of accumulating good flying genes in the population, what do we see? We see a population of very good flyers. Good in all sorts of subtle details including anti-stalling devices; including sensitive nervous control of the muscles that adjust wing shape to every little detail of wind eddies and up-draughts; including more efficient wing muscles that get just that little bit less tired. Wings and tails have evolved to become just the right shape and size, just right in every detail, just right as if a human engineer had perfected the design on a drawing board and tested it in a wind tunnel.

The end products of human design and evolutionary design are both so good, both fly so well, that we find it convenient to forget how different are the two processes of improvement. This forgetting shows itself in the language we use. You will have noticed in this book that I have used a kind of short-hand language. I've written as though birds and bats, pterosaurs and insects set about solving the problems of flight in the same kind of way as human engineers: as though birds themselves solve the problems rather than Darwinian natural selection. The short-hand is convenient partly because it really is short: it takes fewer words than spelling out how natural selection works every time. But it's also convenient because you and I are human and we know as humans what it's like to be faced with a problem, and what it's like to imagine solutions to the problem.

It's tempting to suggest that the similarity between evolution and human design goes further. We might suspect that engineers' new ideas, say for an anti-stalling device, are rather like mutations. These 'idea mutations' are then subject to something like natural selection. An idea may die immediately, when the inventor quickly realises it won't work. Or it may die as a prototype which is seen to fail in a preliminary test, or perhaps in a computer simulation or a wind tunnel, and is consequently rejected. Failing in a wind tunnel is relatively harmless: nobody dies. Natural selection of animal flyers is more cruel: failure really does mean death. Not necessarily a fatal crash, but perhaps the defective design is slower to escape a predator. Or less adept at catching prey on the wing, which increases the chance of starvation. Evolution doesn't have a gentle substitute for death, like trial by wind tunnel. Failure really does mean failure: death or at least failure to reproduce.

Actually, on second thoughts, I've just remembered that young birds of many species can be seen practising flying – we could see it as a kind of playing – before they finally launch themselves into the air in earnest. Maybe that's the bird equivalent of wind tunnel trials: non-fatal trial and error, not just strengthening the wing muscles but probably improving the young bird's coordination and skill. The young of many species of birds can be seen doing what seems to be practising: jumping restlessly up and down while flapping the wings, no doubt exercising the flight muscles and probably honing flying skills at the same time.

Now here's another difference between evolutionary and engineering design (well, it may be just another aspect of the same difference, see what you think). When engineers think of a new design they are allowed to start afresh with a clean drawing board. Sir Frank Whittle (one of several men credited with inventing the jet engine) was not obliged to take an existing propeller engine and modify it in tiny steps, screw by screw, rivet by rivet. Imagine what a mess the first jet engine would have been if Whittle had had to proceed by a step-by-step, tinkered modification of a propeller engine. But no, he started afresh with a whole new idea and a blank sheet on his drawing board.

PRACTICE MAKES PERFECT ⌐
Snowy owl parents look on (mother larger than father) as their child practises flying.

Evolution isn't like that. Evolution is condemned to modify previous designs step by tiny step. And every step along the way has to survive at least long enough to reproduce.

On the other hand, it doesn't follow that evolution always tinkers with a predecessor organ that happened to have the same purpose. In terms of our analogy, the evolutionary equivalent of Frank Whittle might not have been condemned to modify the propeller engine step by tiny step. He could modify some other part of an existing plane, perhaps a bulge in the wing. But evolution can't go back to square one with a completely clean drawing board like a human engineer. It has to start from some part of an existing, breathing animal. And all the subsequent intermediate stages have to be living, breathing animals surviving at least long enough to reproduce. As an example, we'll see soon that insect wings may have started out as modified solar panels for sunbathing, rather than as rudimentary wings.

There are two schools of thought about how innovations come about in human technology. And this reminds me of two schools of thought in modern evolutionary theory. In human innovation there's the 'lone genius theory'. And there's the 'gradual evolution' theory preferred by my friend Matt Ridley in his book *How Innovation Works*. On the lone genius theory, nobody had the faintest notion about jet propulsion until Sir Frank Whittle burst on the scene. But did you notice above that I was careful to say he was one of several men credited with the invention? Whittle patented the idea in 1930 and first got an engine running (not in a plane) in 1937. The German engineer Hans von Ohain filed a patent in 1936, and the first jet plane to actually fly was the Heinkel He 178 with an Ohain engine. That was in 1939, two years before a Gloster E38/39 plane with a Whittle engine took to the air. When they met after the war, Ohain remarked to Whittle, 'If your government had backed you sooner, the Battle of Britain would never have happened.' It's unclear whether Ohain had seen Whittle's patent. In any case there was a 1921 patent by a French engineer, Maxime Guillaume (which Whittle didn't know about). But the point I want to make here is that neither Whittle nor Ohain, nor even Guillaume, thought of it first. The lone genius theory is wrong. There's a long history of inventions more or less resembling a jet engine. Rockets were used as weapons in tenth-century China. In the Ottoman Empire in 1633 a rocket was even used by a man to fly – briefly. Lagâri Hasan Çelebi was reported to have clung to a '7-winged' rocket powered by gunpowder, and fired himself from the Topkapi Palace over the Bosphorus. At some point during the flight he bailed out, fell into the sea and swam ashore where the Sultan rewarded him with gold for his daring feat.

Ridley goes through example after example – the steam engine, the turbine, vaccination, antibiotics, the flush toilet, the electric light bulb, the computer – debunking the lone genius theory in every case. If you ask Americans who invented the light bulb they'll say Thomas Edison. British people might say it was Joseph Swan. Actually, Ridley points out that at least twenty-one people from various countries could claim to have invented the light bulb. Edison does deserve the credit for painfully developing a product you could actually sell. But rather than being invented by any individual genius, the light bulb *evolved* – not genetically, of course, but from mind to mind. It was gradually perfected, step by laborious step. And of course the evolution hasn't stopped. It has improved over the years since Edison's time and we now have LED bulbs which are superior in every way. Technology evolves step by step; perhaps nowhere more dramatically than in the case of the digital computer, which evolves so fast that next year's better (and cheaper) model comes out almost before this year's model has properly warmed up.

Who invented the plane? The Wright brothers. Well, yes, they may have been the first to lift a human pilot, using powered thrust. But gliders go back a long way. The Wrights knew a lot about gliders, having experimented with them over a long period. You could say they took a glider, tinkered with it over a long period, then added a propeller and an internal combustion engine, and took off with it. But that summary conceals a great deal of expert and patient tinkering. They built a wind tunnel and this must substantially have helped them perfect the details. The very first flight by Orville Wright on 17 December 1903 lasted only 12 seconds and he travelled only 37 metres at 6.8 miles per hour. This is not to take away the honour from him and his brother – it was a terrific achievement (and they were sufficiently slighted at the time by snobbish sceptics who didn't believe they'd done it). But the lone genius theory doesn't fit the case. Planes evolved gradually, with roots in gliders and continuing on through early biplanes to the sleek, fast, graceful airliners of today.

I spoke of a mutant falcon and a mutant starling as surviving better because they were better flyers. But this suggests that the improvement had to wait until the right mutation happened to turn up: a bit like waiting for the right 'lone genius' to turn up. But that's not the way things happen in evolution – just as human innovations don't usually have to wait for a lone genius. It's true that mutation is the ultimate source of new 'ideas' in evolution. But sexual reproduction rearranges them, together with other genes, in lots of different new combinations, which are then presented for natural selection. Genes, like engineers' ideas, are shuffled and recombined before being put to the test. It isn't as simple as waiting for a clever mutation (or a lone genius) to turn up.

WRIGHT BROTHERS

The very first powered flight. Note the 'wing warping', which was the Wrights' ingenious way of controlling their flight surfaces. It is not used in modern planes but birds could be said to use a form of it.

CHAPTER 14

WHAT IS
THE USE OF
HALF A WING?

FLYING DRAGON OF THE FOREST

The vertebrate skeleton offers many ways to stiffen a glide surface. The 'flying lizard' spreads its ribs within a membrane of skin. This one is just coming in to a neat landing low down on the trunk of a distant tree.

Chapter 14

What is the Use of Half a Wing?

There are still some people who don't believe in evolution, in spite of the overwhelming evidence in its favour. They want to believe that bird and bat wings, like plane wings, are produced by deliberate creative design: design by some kind of supernatural master engineer. They're called creationists. You won't find them in proper universities. But there are plenty of them in less educated circles.

One of the favourite arguments of creationists dwells on precisely the point I just made in the previous chapter: evolution has to work by gradual, step-by-step change, tinkering with what's already there rather than going straight to the best solution of the problem. And in the case of wings, the way creationists like to put it is the question that heads this chapter: 'What's the use of half a wing?' Yes, they say, a fully fledged wing is all very fine. But a winged animal would have to evolve from a wingless animal, and why would the intermediate stages be any good: a tenth of a wing, a quarter wing, half a wing, three quarters of a wing? Wouldn't an ancestor with only half a wing crash to the ground, if not fatally at least looking foolish? In evolution, each step up the ladder to a proper wing has to be better than the previous step. There has to be a gradual ramp of improvement. All the intermediate animals with partial wings had to have survived. And they had to have survived better than rivals with slightly smaller partial wings. Surely, the creationists say, the intermediates would fail. Surely there's no gradual ramp of improvement. 'What's the use of half a wing?'

How do scientists answer this challenge? Actually it's childishly easy. Think back to the chapter on parachuting and gliding. Remember the flying squirrels and their Australian marsupial equivalents, the flying phalangers. Remember the colugo with its parachute of skin stretched between its four limbs and its tail. The forests of the world, especially South East Asia, house lots more beautiful gliders like those. Flying lizards or flying 'dragons' (the Latin name, *Draco*, actually means dragon) have a web of skin like a flying squirrel. But it isn't stretched between the outstretched limbs. Instead, the ribs shoot out sideways to support a delicate wing of skin on either side; remember the point about evolution exploiting what is already there rather than starting with a clean drawing board? The same forests are home to 'flying' snakes. They don't have any obvious wing stretched between their ribs (and like all snakes they have no limbs anyway). But the ribs push out enough to flatten the whole body, with some curvature of cross-section like a plane's wing,

FLYING FROG

The 'flying frog' spreads its fingers and
toes and the webbing catches the air.

and this provides enough of a parachute effect, perhaps with
help from the Bernoulli principle. They can glide 30 metres from
one tree to another. Again they're slowly descending all the time
but in a controlled way. They look as though they're swimming
through the air, using the same wavy motion as a snake uses
on the ground, or in water. Then there are gliding frogs in the
same forests. Their membrane is stretched not between their
limbs or ribs, but between the outspread toes of all four legs.
None of those gliders can fly properly like a bird or a bat. Their
flight surfaces are not fully evolved wings. They are more like
parachutes. They prolong the fall. How might they have evolved?

All those parachuting animals live in forests, high in the canopy where the sun hits the leaves which feed the whole forest community. Squirrels scuttle about in those high aerial meadows, occasionally leaping from branch to branch. The squirrel's tail has various uses. They flick it as a signal to other squirrels. Or it helps balance them when running and leaping in the trees. For all I know, they might even use it as an umbrella in the rain. It's a sunshade for desert squirrels. But also, as we saw in Chapter 6, its bushy surface catches the air and helps them leap just that little bit further than they could without it.

Why should that matter? If a squirrel falls short of the branch it is trying to reach it might tumble and get seriously injured. There must be a critical distance which a squirrel is capable of jumping without a tail. Whatever that distance might be, a slightly bushy tail would enable it to jump just that little bit further. How much is 'a little bit'? Even if it's only a few centimetres, that would be enough to give individuals with a slightly bushier tail a slight advantage. And then, somewhere up in the canopy, there'll be another, slightly longer critical distance between branches which a squirrel with a yet even bushier tail can just reach. And so on. A forest presents a complete range of distances between branches. So, however far a squirrel can jump with its present tail, somewhere up in the trees there will always be a branch distance that it could have covered if only its tail had been that little bit bushier or longer. An individual of the next generation which has a slightly improved tail is less likely to fall and more likely to survive and pass on genes for making an improved tail.

You already know, from Chapter 6, where this argument is going. The point is that having a bushy tail is not an all-or-none feature. For any size and any degree of bushiness there must be a jumping distance which is just out of reach: a gap between branches which could be jumped if only the tail was just a tiny bit bigger or bushier. And so, we have a smooth gradient of improvement. Which is just what we need for our evolutionary argument.

A bushy tail is not the same thing as a pair of wings. It's not even a parachute like that of a flying squirrel or colugo. But you can easily see how to continue the argument. Any squirrel might have a little loose skin in its armpits. That loose skin will slightly increase the squirrel's surface area without adding much to its weight. This skin flap will work like the bushy tail but more effectively to slightly increase the distance the squirrel can leap without falling. The forest presents a continuous range of distances between branches. Whatever gap a particular squirrel can jump, the canopy will present some slightly longer gaps which another squirrel can jump because it has a slightly larger area of skin flap. And so we have the beginnings of another smooth gradient of improvement. Which is all we need for our evolutionary argument. The end of the gradient will be a flying squirrel, or

IS THIS HOW BATS
GOT THEIR START?
*The colugo has webbed fingers.
But the webbing is a small part of
the huge patagium. To make a bat
from a colugo you'd just have to
grow the fingers.*

flying phalanger
or colugo, with a
full patagium.

'The end' of the gradient?
Why should it stop there? Flying
squirrels and colugos move their limbs while
parachuting, and this enables them to steer their glide. Wouldn't
it have been just a short further step to move the arms repeatedly
and more vigorously until it became a flapping motion? To begin
with, flapping would only slightly prolong the downward glide.
But then, isn't it almost trivially easy to see how that prolongation
could become indefinite? Gradually, stage by stage. Could this
be something like how bats got their start?

As it happens, there are no useful fossils to tell us how
bats first launched themselves into the air, but it's easy to
imagine a plausible gradient. The patagium of a colugo stretches
mostly between the main limb bones and tail. But it also stretches between the short fingers. Webbed feet are common in water

birds and mammals, like ducks and otters. Even some humans are born with short webbing between their fingers. It happens easily because of a peculiar fact of embryology, a phenomenon called *apoptosis* or 'programmed cell death'. In developing embryos, including human embryos, fingers start life webbed, and are carved away from each other like a sculpture. Cells die in a carefully programmed way. Programmed cell death is one of the tricks used to sculpt the embryo. All mammals have webbed fingers while in the womb, before the cells of the webbing died away. Except in otters and other water creatures who need webbing for swimming. Plus... bats, who need it for flying. And a few individual humans, as I said, in whom apoptosis didn't go quite far enough.

Colugo fingers are short. You can easily see how something like a colugo ancestor could gradually lengthen its webbed fingers over evolutionary time, to make a bat. Colugos are loners in the family tree, not closely related to any other mammals. Their closest living relatives after primates are bats. Even if they

CARVING OUT
THE FINGERS
We all had webbing between our fingers in the womb. And some people never quite lost it.

weren't related to bats, the argument I have put would still be a good one. Far from being difficult, for the ancestors of bats, the task of evolving a patagium and then wings would have been easy: just a matter of *refraining* from apoptosis, accompanied by lengthening the finger bones relative to the arm bones. And the selection pressure driving the progression is trivially easy to reconstruct: gradual, centimetre-by-centimetre increase in the leaping distance accompanied by centimetre-by-centimetre lengthening of the webbed fingers to improve sensitive control over the shape of the flight surfaces. Then flapping to improve both control and distance travelled, culminating in true flight.

At this point I must mention that different scientists champion two rival theories of how vertebrates got started on the road to flight. There's the 'trees down' theory and the 'ground up' theory. So far I've mentioned only the 'trees down' theory. I must admit I prefer it. But each theory could be true for different flying animals. For example, bats might have evolved according to the 'trees down' theory and birds by the 'ground up' theory. So let's turn now to the 'ground up' theory, which has indeed been pushed hardest for birds.

Birds evolved from already feathered reptiles that ran on their hind legs. Their ancestors were dinosaurs related to the famously terrifying *Tyrannosaurus*. Running on two legs can be very fast, as ostriches show us today. Now, when you are running fast on your hind legs, your front limbs are not directly involved, unlike those of galloping mammals. But perhaps they could help in some other way. Athletes pump their arms vigorously back and forth while running. Ostriches, which are among the fastest-running land animals, use their 'arms' (or you could call them stubby wings, because they are still recognisably wings, inherited from flying ancestors) for balance, especially when turning.

Maybe reptiles running fast on their hind legs could have run more efficiently, by interspersing leaps as they ran. Like flying fish in water. Feathers, originally evolved for heat insulation purposes, could have assisted the leaps in the same kind of way as I mentioned for the bushy tails of squirrels. The feathers on the tail and the arms, especially, would have prolonged the jumps in the same kind of way as a developing patagium. The arms outstretched for balance purposes would have been especially helpful here and could have developed into rudimentary wings, not enabling true flying yet, but prolonging the leaps. The argument is similar to the one about the continuous range of branch distances presented by trees. However far a reptile could leap without feathered arms, it could leap just that little bit further by spreading them. Peacocks, as we saw earlier, and pheasants are not great flyers. They usually land soon after take-off: peacock flights are little more than prolonged jumps, which serve to get them out of danger, like when a flying fish temporarily takes to the air to escape a pursuing tuna. As the generations passed, there would have been a ramp of steadily longer escape leaps making use of a steadily larger surface area of feathered arm, culminating in true flights of indefinite length.

Turning from prey to predator, we have the 'pouncing predator' theory. According to this idea, a type of feathered dinosaur specialised in ambushing prey. It lurked in some vantage place, perhaps on a steep bank, waiting for prey to pass. It then pounced. Feathered arms and tail kept the predator in the air for a short while, which meant it could pounce from a greater distance. There would have been a gradual ramp of improvement like the one we thought about for the flying squirrels, but in this case it would be a ramp of steadily increased pouncing distance.

And here's another possible variant of the 'ground up' running theory. Insects discovered flying long before any vertebrate, and swarms of flying insects would have been a rich source of food waiting to be exploited by evolving vertebrates. Maybe fast-running reptiles leapt into the air to catch them. They might have snapped their jaws at them like dogs do today. Or, like cats, they might have used their high-reaching arms. Ordinary pet cats can jump as high as 2 metres into the air, and can catch flying birds in their outstretched paws, as well as insects. Big cats like leopards do the same, and catch bigger birds. Could ancestral reptiles have done something similar when going after flying insects? And could rudimentary, non-flying 'wings' have helped?

First, let's take a look at a famous fossil, *Archaeopteryx*. In many respects it was intermediate between birds and the animals we normally think of as reptiles. It had wings much like modern birds, but with protruding fingers. Unlike modern birds, it had teeth like a reptile. Well, I say unlike modern birds but... in *Hen's Teeth and Horse's Toes*, one of his lovely books of natural history, the late Stephen Jay Gould describes how ingenious experimental embryologists succeeded in getting chicken embryos to develop teeth. In the lab they rediscovered an ancestral ability that had been lost for many millions of years. *Archaeopteryx* also had a long, bony reptilian tail, which doubtless was an important flight surface and stabiliser, along with its wings.

It has been suggested that the ancestors of *Archaeopteryx* found their feathers (originally evolved for heat insulation) useful for catching insects. The feathers on the arms grew larger as a kind of butterfly net to sweep up flying insects. And, as it turned out, a butterfly net of feathers had the additional benefit of working as a crude flight surface. Not yet real flight, but the feathered arms could have helped the leaping reptile reach higher-flying insects, as well as helping to scoop them in. A flight surface needs a large area and so does a butterfly net. When leaping into the air to catch an insect, the 'butterfly net' served as a crude wing,

---

IS IT A BIRD? IS IT A REPTILE? WHO CARES? ↠
Archaeopteryx is somewhere close to the reptilian ancestor of all birds and is therefore intermediate. It had teeth, prominent fingers and a long stabilising tail.

which extended the length of the jump and raised its height. The sweeping motion of the wing when scooping up an insect would have looked a bit like the flapping of a wing, and this could have provided additional lift. Gradually the arms lost their 'butterfly net' function as it became replaced by the wing function. And so, according to this theory, true flapping flight in birds evolved. I must say I find the 'butterfly net' theory, and the other 'ground up' theories less plausible than the 'trees down' theory, but I mention them for completeness because some biologists favour them.

Yet another version of the 'ground up' theory is the 'running up a slope' theory. Ground-dwelling animals often scuttle up trees, for example, to escape predators. Squirrels spring to mind, but many other animals do it too, if less expertly. Not all tree trunks are vertical. Some fallen dead trees, or big branches that have broken off, provide a slope. Indeed, you can find a complete range of angles, from horizontal to vertical. Now, imagine trying to run up a 45 degree slope. You could assist your climb by flapping your feathered arms. Not yet wings, not yet developed enough to glide through the air, they nevertheless, when flapping up a sloping tree trunk, could provide just that little bit of added lift and stability that make a difference. Now, yet again, we have a gradient of improvement, literally as well as metaphorically, a gradient. And at the same time as the proto-wings were developing for a 45 degree slope, they would automatically be available for improvement ready to tackle a 50 degree slope.

And so on. That all sounds a bit speculative, but some beautiful experiments have been done on Australian brush turkeys.

☞ Not that it matters, **by the way**, these birds are not really turkeys. They're called turkeys because they're the nearest Australian approach to something that looks like an American turkey. They are 'megapodes', birds that have evolved a remarkable method of incubating their eggs. They don't sit on them. Instead they build a large compost heap and bury their eggs in it. Bacteria in the rotting compost generate heat, which serves to incubate the eggs. Incubated eggs are fussy about temperature. When parents sit on eggs the temperature is exactly right: the parent's accurately regulated body temperature. So how do megapodes regulate the temperature of their compost heap? By taking plant material off the top of the compost heap when it's too hot, and adding some, like a blanket, when it's too cold. The bill has evolved to become a thermometer and they stick it in the compost heap to measure the temperature.

I couldn't resist that little digression about megapode compost heaps. I find them fascinating. But for this book what matters is that the megapode chicks are extremely capable and independent when they hatch. They have to be, as their parents are not around to look after them. Remarkably, they are even able to fly the day after they hatch. But flight is not their preferred way to get away from predators. Instead, they run up tree trunks. And as they do so, they flap their wings to help them up the slope. They can even climb vertical tree trunks using their flapping wings to help them. You can easily see that, where full flapping wings can assist today's megapode chicks in climbing a vertical surface, less well-developed wings could have helped their ancestors to climb shallower slopes. And the wings would have been effective only if flapped – as the brush turkey chicks flap today. Once again, we have a gradient of improvement (a gradient of gradients, as it happens). And gradients, of course, are what we need if we are to explain, 'What is the use of half a wing?' The fact is that, contrary to the creationists, it isn't at all difficult to see lots of ways in which flight could have evolved gradually, step by step. Lots of ways in which half a wing would be better than no wing.

How about insects, which discovered flying hundreds of millions of years before vertebrates; how did that come about? Most insects today have wings; though some, like fleas, have lost them, being descended from winged ancestors. They are called 'secondarily wingless'. As we've already seen, worker ants and termites are descended not just from winged ancestors but from winged parents, for queens and males have wings. There are also some primitively wingless insects, such as silverfish and springtails, whose ancestors never had wings.

As with all arthropods (insects, crustaceans, centipedes, spiders, scorpions, etc.), the insect body plan is segmented.

Segmentation is more clearly seen in centipedes and millipedes. They are built like a train with lots of trucks arranged in a line, almost all the same as each other, each segment bearing legs. In other arthropods, like lobsters and insects, the segmentation is still there but it's more complicated: the different segments ('trucks') have evolved to become different from each other. Trains, too, sometimes have lots of identical trucks, sometimes the trucks share little except similar wheels and identical coupling mechanisms. We vertebrates, too, are segmented; the vertebral column makes this obvious. But even our head turns out to be segmented if you look at it carefully, especially in the embryo.

In insects the first six segments constitute the head but are squashed up together so their train-like arrangement is obscured, as it is in mammals. The next three segments are the thorax. The rest of the segments make up the abdomen. All three segments of the thorax have a pair of legs, and in most insects the last two thoracic segments have wings as well. Flies (and their relatives like mosquitoes and gnats) are a special case as we've already seen: they have only one pair of wings, the second pair having shrunk in evolution to become haltere 'gyroscopes'.

Unlike vertebrate wings, insect wings are not modified limbs. As we've already seen, they are extensions of the thorax wall. All six legs remain free for walking. There are various theories of how the wings originated. Many flying insects have a juvenile stage, which lives in water, and emerges into the air when adult.

Some of these juvenile stages, *nymphs*, have gills for breathing underwater. Unlike fish gills but, coincidentally, like tadpole gills, these nymph gills are feathery outgrowths. Some scientists think insect wings evolved from modified gills. Another theory is that aquatic nymphs developed 'sails' for scooting across the surface of the water, and these later became wings.

A currently fashionable theory is that small stubs – flanges, extensions of the thorax – stuck out as sunbathing surfaces, 'solar panels', to warm the body up, rather than as flight surfaces. The authors of this theory did experiments with model insects, partly in wind tunnels. Their results suggest that very small thoracic flanges are less good for aerodynamic purposes than they are for soaking up the sun. Larger wing stubs become better aerodynamically. Where flat projections from the thorax are concerned, there's a threshold size for flight surface to overtake solar panel as the main advantage. So, if stubs started off as sun absorbers, the insects only had to get bigger, which is often and easily done for lots of reasons. As the wings became larger, they found themselves automatically becoming more useful as flight surfaces. These later evolved into proper wings.

So, according to this theory, the initial steps up the evolutionary gradient were taken for reasons of solar heating. It's obvious that that would be a smooth gradient: the larger the stub area, the more sun ray absorption. And when the threshold size was exceeded, the stubs automatically became useful, first for gliding, later for flapping, using muscles already present in the thorax. Remember from Chapter 8 that insect wing flapping is normally achieved by muscles that simply deform the thorax. And reflect that the best sunbathing panel is likely to be thin – like a wing. Gradual, step-by-step increase in body size incidentally pushed thoracic flanges over the threshold and they automatically became more useful as flight surfaces.

Whatever particular theory you favour out of the many on offer, we again conclude that 'What's the use of half a wing?' is not a problem. In insects as well as pterosaurs, bats and birds, gradual, step-by-step evolution by natural selection takes care of it.

NOT EVEN HALF A WING
The flying snake of the forest shows how it can be done, simply by flattening the body to double its width, and 'swimming' sinuously through the air from one tree to another.

CHAPTER 15

THE OUTWARD URGE: BEYOND FLYING

A SCENE ON MARS AT THE TIME OF WRITING

Will the future see a flourishing colony of humanity? It won't be easy to phone home. Each word will take between three and 22 minutes to arrive, depending on relative orbital positions.

CHAPTER 15

THE OUTWARD URGE:
BEYOND FLYING

I began the book by asking whether you – like me – ever dream of flying like a bird. But now, in closing, I wonder whether you have a dream that one day you might leave our home planet altogether and fly all the way to Mars? Or one of the moons of Jupiter? Or Saturn? When I was young, that dream belonged to science fiction. I was fond of a comic strip character called Dan Dare, Pilot of the Future. He and his Lancashire sidekick Digby would spring casually into their spacecraft, seize the joystick and zoom off in the general direction of Jupiter.

Nowadays we know it isn't as simple as that. It takes several years to get there. It's a massive, cooperative project involving hundreds of engineers and scientists, who compute orbits well in advance and plan a complicated schedule of gravitational slingshots around other planets on the way. Even travelling to Mars takes months. But it is a real possibility. Unmanned spacecraft have already done it. Elon Musk not only wants to send his rockets to Mars. He wants to set up a colony there. And he has a serious reason.

You remember the discussion in Chapter 11? The mathematical theory of why animals and plants have evolved an urge to send at least some offspring out to try their luck at making a living far away? Even if the parents themselves live in the best possible place? The essential reason, you'll remember, was that sooner or later a catastrophe such as a fire or flood or earthquake will hit wherever you are, so even the best place in the world will cease to be the best place after all, quite the reverse.

Well, Earth is certainly now the best place for humans to live. And Mars is a terrible place to live. But could Earth one day be hit by a catastrophe so bad that the only way humanity could survive would be in the form of a colony of pioneers elsewhere? What sort of catastrophe? There are several possibilities including the long-term effects of climate change, a lethal pandemic and various kinds of high-tech warfare, including biological warfare, getting out of hand. But I'll mention one further possibility to stand for all of them. Admittedly it's the most unlikely one in the short term but I'll talk about it because it is probably the one least likely to have entered the minds of most people. And, although unlikely in the short term, it will for sure eventually happen. And when it does, it will be worse than anyone's darkest nightmares. If we are to avoid it, the remedy will have to be one that pushes our flying arts well beyond anything we have covered in this book so far.

You know what happened to the dinosaurs. Their kind had ruled the land for 175 million years. For the dinosaurs, in their different ways, Earth was the perfect planet until... out of a clear blue sky, with no warning, a mountain-sized chunk of rock hurtled at 40,000 miles per hour straight into what is now the Yucatan peninsula in Mexico. The local dinosaurs were instantaneously vaporised by the searing heat at more than 2,000 degrees Celsius. But that was just the beginning. The impact was equivalent to several billion Hiroshima-sized atom bombs exploding simultaneously at the same spot. The sea boiled and a mile-high tidal wave raced off around the world. But, finally, it probably wasn't the explosive heat or the forest fires or the tsunami that killed the last of the dinosaurs. The shattering impact threw up dense clouds of ash, dust and sulphuric acid droplets which darkened and cooled the world for years. The Yucatan dinosaurs were the lucky ones. They perished instantly. Those living further away had a more drawn-out fate: cold starvation as the plants on which they depended died for want of sunlight. We mammals survived by the skin of our tiny teeth, probably by hibernating underground. Eventually we emerged, twitching our whiskers and blinking, bewildered in the slowly returning sunlight. And now here we are, descendants of those few survivors: we who evolved into mice and rhinos, elephants and kangaroos, antelopes, whales, bats and humans. But we were very lucky. And we may not be so lucky next time.

For it will happen again. Lesser meteors hit Earth often, and it's only a matter of time before we are hit by another one as big as the dinosaur-killer of 65 million years ago. Or even bigger. Don't lie awake worrying about this. Although it could happen in your lifetime, or even next week, that isn't very likely: 65 million years is a long time and we could go for as long again without a major impact. Nevertheless, some people, including me in my more pessimistic moments, think it is time for humans to start preparing for the possibility. Nobody else will. Our planet relies on us.

One way to prepare is to develop the technology to detect, intercept and deflect an incoming projectile whose elliptical orbit around the sun threatens to meet our near circular one. We are not too far from knowing how to do this. A major step in the right direction was the *Rosetta* spacecraft's feat of making a landing on a comet. The next step would be to nudge the threatening asteroid or comet into a slightly different orbit around the sun. Speed it up a little, or slow it down a little so that its orbit no longer intersects ours. The necessary change in speed, either way, is surprisingly small. But we'd need to exert a very large force to influence the sort of mountain-sized meteor that could threaten our survival.

But whatever the threat to Earth, whether it's a comet or an unstoppable plague, there's something to be said for heeding the lesson of Chapter 11 and founding a colony of humans on another planet such as Mars. Of course, Mars might also be struck by a giant asteroid. But both planets would not be struck by the same one – or by the same plague – and you've surely heard the proverb about putting all your eggs in one basket. Founding a colony on Mars would be hugely difficult – there's no oxygen worth speaking of and little water. It wouldn't save the vast majority of humans as individuals. But it could save our species. There would at least be a memory, an archive of all that we have achieved over the centuries: the music, the art and architecture, the literature, the science. And there would be the possibility of eventually recolonising Earth and starting afresh. That, at any rate, is one reason for wanting to go to Mars.

In Chapter 11, about the outward urge of animals and plants to send offspring far from present comfort into the wild unknown, were you reminded of anything in human history? A spirit of adventure? Of pioneering recklessness? Did you think of the urge that drove the great explorers like Christopher Columbus, who sailed west to America without having a clue where he was going? Or Ferdinand Magellan, whose expedition sailed right round the world (although he was killed before reaching home)? They were followed by would-be colonists seeking refuge, at least in the case of America, from persecution, but not knowing what hazards awaited them.

Earlier, Vikings led by Eric the Red were driven by a similar outward urge to sail into the westerly unknown and set up home in Greenland. Eric's son Leif Ericson went further and reached North America half a millennium before Columbus. Nobody knows when the ancestors of today's Native Americans crossed the frozen Bering Straits from Asia, but who would confidently deny that they too were led by the same spirit of adventure? Eric the Red's Viking dynasty of westering adventurers may have inspired John Wyndham, the science fiction author, when he wrote *The Outward Urge*, whose title I have borrowed for this chapter. His heroes were seven generations of a family whose inherited urge to explore the wild unknown led them ever deeper into space.

I write these last paragraphs in my hotel room in the city of Zurich, where I am attending an inspirational conference: STARMUS, a meeting of scientists, rock musicians and astronauts to commemorate fifty years since men first walked on the moon. Many of the astronauts here are veterans of the US Apollo programme. Some of them walked on the moon. One by one they have stood up at the conference and eloquently told how the experience of going into space, walking on the moon, floating weightless, seeing Earth from the outside against a soot-black sky, changed them. They are mostly drawn from the ranks of fighter test pilots. Fighter pilots are not, on the whole, known as natural poets, not typically emotional, and this makes their testimony the more moving. I see them as heirs of the great seafaring explorers, the Ericsons, Magellans, Drakes and Columbuses of past centuries. Or, perhaps even more poignantly, the Polynesians whose canoes colonised island after island in the vast Pacific, even penetrating as far as the remote Easter Island – a journey that perhaps seemed to them like venturing to the moon does to us.

HOW DID THEY EVER DISCOVER EASTER ISLAND? *Does the adventurous spirit of the Polynesian island voyagers live on in the 'outward urge' of our species to colonise Mars and, one distant day perhaps, reach for the stars?*

But also, since I am an evolutionary biologist, I cannot help thinking too of the deeper past. Of our ancestors who, a thousand centuries ago, walked out of Africa and colonised Asia, Europe, Australia and – over the Bering Straits – became the first true Americans. Were they too driven by the same outward urge? Or did they just wander, generation by generation, never dreaming they were part of a great, historic exodus?

Or, to revert to a timescale of millions of centuries, was it the same outward urge that led the first fish to venture onto land? Was it an unusually adventurous, enterprising lobefin? Or just a fortuitous accident? And what of the first reptile to take to the air? The first feathered dinosaur whose leaping ambition was destined to birth the great family of birds. A brilliant, pioneering individualist? Or pure happenstance? I'm genuinely curious to know.

Back to the Zurich conference – the other half of the gathering consists of scientists, including several Nobel prize-winners, the mental counterparts of astronauts taking those first tentative, physical steps into the weightless unknown. The liberation from gravity which began with the insects, birds, bats and pterosaurs and continued with the balloonists and aviators of our own species has culminated, literally, in the weightlessness of the astronauts and, symbolically, in the imaginative flights of fancy of the scientists.

And from my pillow, looking forth by light
Of moon or favouring stars, I could behold
The antechapel where the statue stood
Of Newton with his prism and silent face,
The marble index of a mind forever
Voyaging through strange seas of thought, alone.

William Wordsworth,
from *The Prelude*, 1799

Wordsworth's lines about Isaac Newton might even more appropriately have referred to Stephen Hawking, who, cruelly prevented from moving physically, voyaged through strange seas of thought, alone behind his forever-silent face. I think it appropriate that at the Zurich conference the Stephen Hawking Medal was presented to the visionary engineer and 'outward urge' prophet to whom this book is dedicated.

I think of science itself as an epic flight into the unknown: whether it's a literal migration to another world, or whether it's a flight of the mind, soaring abstractly through strange mathematical spaces. Perhaps it's leaping upwards through a telescope to distant, retreating galaxies; or diving through a glowing microscope tube, deep into the engine-rooms of the living cell; or racing the particles round the giant circle of the Large Hadron Collider. Or maybe it's flying through time, either forwards in the company of the majestically expanding universe, or back through the rocks beyond the birth of the solar system, towards the origin of time itself.

Just as flying is an escape from gravity into the third dimension, so science is an escape from the mundane normality of the everyday, spiralling up through rarefied heights of the imagination.

Come, let's spread our wings and see where they may take us.

ABOUT THE AUTHOR

RICHARD DAWKINS

Richard Dawkins was the Inaugural Charles Simonyi Professor
of the Public Understanding of Science at Oxford University.

His books have sold many millions of
copies and have been published
in more than forty languages.
They include *The Selfish Gene*,
The Blind Watchmaker, *The
God Delusion*, *The Magic
of Reality* and a string of
other bestsellers. In 2017, to
celebrate the 30th anniversary
of its Science Book Prize, the
Royal Society conducted a poll
to identify 'the most inspiring science
books of all time': *The Selfish Gene* topped the poll. Richard
Dawkins has honorary doctorates in both science and literature,
and is a Fellow of both the Royal Society and the Royal Society
of Literature. He has presented science documentaries on both
the BBC and Channel 4, and was selected to give the 1991 Royal
Institution Christmas Lectures for Children, televised by the
BBC. In 2013 he was voted the world's top thinker, in *Prospect
Magazine*'s poll of 10,000 readers from over 100 countries.

Jana Lenzová

Jana Lenzová, born and raised in Bratislava, Slovakia, is an illustrator, translator and simultaneous interpreter. Her two great passions are languages and drawing. The former led to the latter. After she had been commissioned to translate Richard Dawkins's *The God Delusion* into Slovak, she began contributing to his books as an illustrator. Jana also illustrated several book covers and her artwork has been featured on various blogs, including the CBC/Radio-Canada blog covering the 2014 Winter Olympics.

Jana works digitally, from her initial sketches to building up the colour on her illustrations. Below are some of the illustration stages for the hummingbird – the very first image Jana created for this book.

ACKNOWLEDGEMENTS

With thanks to Anthony Cheetham, Georgina Blackwell, Jessie Price, Clémence Jacquinet, Steven and David Balbus, Andrew Pattrick, David Norman, Michael, Sarah and Kate Kettlewell, Greg Stikeleather, Lawrence Krauss, Leonard Tramiel, Jane Sefc, Sonjie Kennington, Henry Bennet-Clark, Connie O'Gormley and the late, much missed Rand Russell.

Black and white images © Shutterstock
Contributor details below.

Index